Macmillan Computer Science Series

Consulting Editor
Professor F. H. Sumner, University of Manchester

Barry Morrell and Peter Whittle, *CP/M 80 Programmer's Guide*
Derrick Morris, *System Programming Based on the PDP11*
Pim Oets, *MS-DOS and PC-DOS—A Practical Guide*
Christian Queinnec, *LISP*

(continued overleaf)

W. P. Salman, O. Tisserand and B. Toulout, *FORTH*

L. E. Scales, *Introduction to Non-linear Optimization*

Peter S. Sell, *Expert Systems—A Practical Introduction*

Colin J. Theaker and Graham R. Brookes, *A Practical Course on Operating Systems*

J-M. Trio, *8086–8088 Architecture and Programming*

M. J. Usher, *Information Theory for Information Technologists*

B. S. Walker, *Understanding Microprocessors*

Peter J. L. Wallis, *Portable Programming*

Colin Walls, *Programming Dedicated Microprocessors*

I. R. Wilson and A. M. Addyman, *A Practical Introduction to Pascal—with BS6192, second edition*

Introduction to Discrete Mathematics for Software Engineering

Tim Denvir

Praxis Systems plc

MACMILLAN

First published 1986

Published by
MACMILLAN EDUCATION LTD
Houndmills, Basingstoke, Hampshire RG21 2XS
and London
Companies and representatives
throughout the world

Printed in Great Britain by
Camelot Press Ltd, Southampton

British Library Cataloguing in Publication Data
Denvir, Tim
    Introduction to discrete mathematics for
    software engineering.—(Macmillan computer
    science series)
    1. Electronic data processing—Mathematics
    510   QA76.9.M35

ISBN 0–333–40736–9
ISBN 0–333–40737–7 Pbk

Contents

To Brenda, James and Eleanor;
to my colleagues at Praxis Systems plc;
and to the memories of Alan H. Day and A. Richard Tate

Foreword

Notations from areas of discrete mathematics can be used to record the specifications of software systems; reasoning based on the properties of such notation can be used to show that designs satisfy such specifications. The sufficiency of discrete mathematics shows that, in some ways, it should be easier to verify software designs than to ensure safety in those engineering disciplines which have to cope with continuous variation, stress, decay, etc. In other ways, software design is made difficult by the inability to increase safety by adding redundancy.

The correctness of a software system must be established by mathematical proof. There is a growing acceptance that 'formal methods' are essential in achieving computer systems which are safe in use and economical in creation and maintenance costs.

There already exist a number of books which show how formal methods can be applied to software development. Most of these books assume a prior knowledge of discrete mathematics. There are, of course, many books which teach the relevant mathematical ideas, but these are not related to software problems.

Tim Denvir's book bridges the gap. This is a book with well-chosen examples which takes the reader from the elementary details of set and logic notation to their use in specification and proof. The book has evolved from industrial courses and should provide the sort of 'new-readers-start-here' material which will enable many more software engineers to begin to apply formal methods.

Marktoberdorf   Cliff Jones
August 1986

Preface

The title of this book has been chosen with some deliberation. The subject of the book is *Discrete Mathematics*. It does not aim to teach any techniques or methods of software engineering *per se*, but rather the mathematics which lies at their foundations. It is an *Introduction* only, because the elements of the subject are all that are required for this purpose. Finally, it is about Discrete Mathematics for the activity of *Software Engineering*, whether that activity is carried out by software engineers, accountants, civil engineers, astronomers, clerks or anyone else. Thus the title is intended to invite all who propose to 'engineer software', whatever their principal occupation.

This book has four primary aims: to teach to a depth adequate for its appreciation and practice the mathematical theory which underpins a scientific approach to software engineering; to impart sufficient skills in Discrete Mathematics to enable readers to practise software engineering in a scientific, disciplined manner; to convince readers of the desirability and applicability of Discrete Mathematics to software engineering; and to provide a cognitive stepping stone such that readers will be able to deal with more advanced works in computer science and related topics, as found in the ever-increasing range of scientific journals and books.

To support these aims, examples are drawn from software engineering which model the discrete mathematical concepts. These examples are thematically developed and elaborated as further concepts are introduced. To develop readers' skills, exercises are also presented. Some of these are formalistic and provide practice in the notational symbolism. Others apply the mathematical concepts to practical domains such as data processing or telecommunications. Some of the exercises are designated 'workshop'. These are intended for supervised group discussion and solution.

One of the workshop exercises concerns a control system for a nuclear power station. As I write this preface, the tragic explosion at Chernobyl has just occurred. This particular choice of workshop exercise may therefore be considered an unhappy one. While I believe that a scientific, mathematically based approach can only increase the safety of any piece of engineering, alone it cannot guarantee it; many other principles and practices need to be applied, and some risks will almost certainly remain.

This book is relevant for courses in discrete mathematics which form part of an undergraduate curriculum in computer science or software engineering. It can also be used to support courses for software engineers in industry. In fact, it was out of such an industrial course in STC plc that this book had its origins. Practising software engineers found that a grounding in discrete mathematics was of substantial advantage when

trying to apply techniques such as VDM (Jones, 1980, 1986), CCS
(Milner, 1980) or Dijkstra's 'discipline of programming' (Dijkstra, 1976).
This book thus owes a debt to over 120 students from the ranks of
STC plc who attended the courses and helped the author and other
lecturers to remedy its initial deficiencies.

Since the initial motivation provided by the course, the production of
this book has struggled up a long and stony path. Over this time the
support and encouragement of many colleagues has given my fragile
enthusiasm and determination the impetus needed to continue the work. I
am especially grateful to Paul Taylor, who, as a colleague at STL took
part in giving the Discrete Mathematics course to STC engineers, and
later when at Brighton Polytechnic read the manuscript and suggested
many improvements to the quality of my proofs of theorems and other
mathematical presentations. Several of the illustrative examples are his. I
am also particularly grateful to Cliff Jones who gave me encouragement at
a critical time and made helpful suggestions on my final draft.
A. McGettric, J. Van Leuwen and a number of anonymous reviewers
made substantial comments on the first two drafts, and Chris George,
Roger Shaw and Mel Jackson supported me in giving the STC course
which catalysed the work. Kathy Evans persevered with my handwriting
in producing the whole of the first draft, and Standard
Telecommunication Laboratories kindly allowed me the use of their
facilities for preparing the various drafts.

Any serious attempt to communicate ideas always reveals to the
communicator new insights into the subject matter. When I started
writing this book I thought it would take less than a year, but it has in
fact taken more than four. During that time my motives, while always
being to communicate, have shifted their focus. I started by trying to
communicate a skill, but now I hope I may communicate a further kind of
understanding. Computer programs are built of abstractions at all levels.
They are like poems whose language is pure thought, whose form is of
science, and whose power, if controlled by an engineering discipline, can
be put to extending ourselves and our environment or destroying them.

Mathematics is perhaps the most abstract form of thought the human
mind can contrive. The pleasure I derive from mathematics has always
disproportionately exceeded my skill in manipulating it. At the end of
these four years I now hope that what I have produced will kindle in
others, not merely a skill but more essentially, an understanding of the joy
to be found in mathematics, and with that, a perception which will stir
software engineers to direct their skills to the service of humanity and our
environment and not to their degradation.

TIM DENVIR

London
April 1986

Chapter 1 Introduction

This book deals with Discrete Mathematics as an approach to formulating problems in such a way that their solution can be contemplated. It attempts to show how problems can be stated in terms of discrete mathematics so that it becomes possible to manipulate these problem statements, infer deductions from them, and provide arguments that constructed solutions do indeed solve the original problems.

The problems in question are those typically found in the field of software engineering: in information systems, in embedded computer applications such as digital process control, communication systems, navigation and user–machine interfaces, in commercial data processing such as personnel records and stock control, and in an ever growing number of other applications.

The discrete mathematics treated in this book is at an elementary and summary level from the point of view of a mathematician. There is a compensating depth of application of the fundamental mathematical theory to a range of software engineering examples. Hence this book is written for software engineers or those aspiring to be software engineers, who may perhaps be following a course in Computer Science.

This book is therefore suitable as a course book for the Discrete Mathematics component of a computer science undergraduate curriculum, especially where such a curriculum is oriented to software engineering. It is also suitable for courses on software engineering in industry, designed perhaps for those who are transferring to software from some other engineering discipline, or for those who have been in the industry for some years and require updating on aspects of theoretical foundations and approaches.

1.1 SOFTWARE ENGINEERING

The Craft

Engineering disciplines in general evolve through a number of recognisable phases. The first phase is that of a craft, where adept people learn to be experts through a period of apprenticeship and exposure to the methods of work of existing experts. There is little or no formal training

1

at this stage. The acceptability of one artifact over another produced by the apprentice, provided of course that they function to an adequate standard, is judged by ill-defined criteria: the master craftsman pronounces one product of his trainee as 'elegant', 'well-structured', 'a good piece of engineering', another the converse. In this evolutionary phase of the discipline, the industry does produce artifacts which work, some of them indeed becoming long lasting monuments to the early craftsmen. In general, however, the standard of reliability or efficiency of these artifacts is erratic: the early steam engines were inefficient, early electrical engineering devices were over-engineered and cumbersome. The early compilers took many years of effort to craft on an individual basis, and some of the techniques first used for particular problems such as expression evaluation could, one can imagine, be eventually forgotten now that parser generators are common practice.

The 'early software artifacts' had several major disadvantages. Firstly, one could always expect to uncover another hidden error after years of use, for there was no means of establishing that a program was correct or likely to be correct. Secondly, in the absence of systematic techniques of construction, large programs required what would now be regarded as excessive effort to produce. Thirdly, because of the preceding two factors, products beyond a certain critical size and complexity simply could not have been constructed twenty years ago.

There is a danger however in becoming too patronising about the 'early' software engineering enterprises and methodology. Much of the same primitive approach to constructing software has re-emerged as a result of the sudden increase in demand for programming produced by the many microcomputers now on the market. This primitive approach is typified by the programmer who constructs a program by first writing it and then tinkering with it until a number of arbitrarily chosen tests succeed.

One mission of this book is to assist the evolution of software engineering by giving its practitioners an extra nudge, indeed if possible a thrust, out of the era of craftsmanship and into the dawning age of software as an engineering discipline!

The Science

Engineering and science are linked. The establishment of a scientific theory which stands up to experimental tests and observation impacts engineering practice by providing a rationale for the techniques which are successful. For example, the theory of thermodynamics provides an explanation for the functioning of steam engines, which permitted more efficient designs. Electromagnetic theory provides a basis for the rules laid down by the electrical engineering standards authorities for safe wiring regulations.

Any science requires some form of mathematics for the formulation of its theories. Indeed many branches of mathematics have developed as a result of the stimulus provided by some particular science. In any branch of mathematics one may formulate premises, and from them infer conclusions according to certain inference rules. When used to express a scientific theory, the premises represent the laws of the science, and the conclusions which can be drawn from them are the predictions which can then be tested by experimental observations.

The Discipline

The way in which this interacts with engineering practice is that the predictive ability of the mathematical modelling inherent in the science enables one to reason about the behaviour of a design before it is built. This means that the process of design ceases to be a hit and miss affair; there is no need for one to approach the task by a process of 'construct and tinker until it appears to work'. There is the opportunity of examining and comparing different designs, establishing whether or not they will meet their specifications, and comparing their performance in advance. In this way the mathematical modelling inherent in the science provides the basis of an engineering discipline.

We referred to specifications. To specify the behaviour or performance of an engineering artifact in anything more than an entirely vague manner, we need to be quantitative in some respect; if we are not to be satisfied with phrases like 'reasonably accurate', 'a high speed', 'adequate efficiency', all of which are open to interpretation, we need to quantify these criteria in order to provide a specification against which a design may be judged and an actual constructed prototype can be tested. Quantification is again an aspect of mathematical modelling: we take a physical or engineering observation such as the frequency of piston cycles in a steam engine and relate it to a mathematical entity, namely a number and a dimension. The predictive power of the mathematics underlying the scientific theory enables one to determine in advance whether the design will meet the specification.

Software as an Engineering Discipline

Computer Science is the study of the theories of computational machines and the mathematical modelling associated with those machines. This science has developed by considerable strides in the last fifteen years in terms of the theory of programs. A mathematical formulation of the syntax or the form of programming languages was developed in the late 1950s from work on natural languages by Chomsky and others (Chomsky, 1956). These schemes give a way of defining the rules for constructing syntactically correct programs for any given language, but they say nothing about the meaning or 'semantics' of a program. Theories giving mathematical models of program language semantics were advanced in the late 1970s: the seminal work of Hoare (1969) and Scott and Strachey (see Stoy, 1977 and Scott, 1970 for example) are notable landmarks in this development. One may wonder what the semantics of programming languages has to do with a discipline of software engineering. The fact is that a theory of program language semantics is a means of formulating the behaviour of any program, and thus it provides a basis for reasoning about a program and drawing inferences about its behaviour. One may therefore determine if a particular program fulfils a specification: the specification will lay down constraints about the behaviour of the program, and quite probably its exact requirements. By having a means of predicting a program's behaviour expressed in mathematical terms, we thereby have a way of proving in advance its adherence to its specification.

All this is looking far ahead however. Program language semantics and proving that a program meets a specification is dealt with in the very last chapter. The reader is progressively led along a path to that goal in the earlier chapters, in which the mathematical formalisms necessary for expressing specifications are developed and explained.

Some Principles

Programming is a crucial creative activity in the process of software engineering. It is however a design process; the programmer decides the exact construction of the final artifact. Programmers are not like wiring operatives or bricklayers who simply construct something according to another's design. They make the final decisions about what algorithms are to be used in order to achieve the desired effect.

Programmers therefore have the task of constructing an artifact which must meet a given specification. It is now widely contended that the only language adequate for expressing the specification for a program is a language based on Discrete Mathematics, and the only way of defending the assertion that a particular program meets that specification ultimately relies on techniques of mathematical proof. Such proofs require predicate logic and a scheme of relating the meaning of programs to their specifications.

Readers should not be alarmed by these statements. This book works through the necessary ideas in small progressive steps which are illustrated by many small examples. However, the philosophy of this book is firstly that software engineering is essentially a mathematical activity, in particular an activity of discrete mathematics, and secondly that the discrete mathematics is actually not difficult.

1.2 KNOWLEDGE REQUIRED TO READ THIS BOOK

The knowledge required to read this book is in two categories. Some familiarity with the ideas in high level programming languages is needed, and some preliminary knowledge of mathematics.

Readers will need to be familiar with the elements and usage of some fundamental high level language constructs. These broadly consist of variables, expressions, assignment statements, loops, conditions, functions, procedures, arrays, and at least elementary data types.

Also knowledge of some elementary mathematics is required, although nothing much more advanced than the British O-level standard is deliberately assumed. This chapter (1.3) summarises some of these elementary concepts.

1.3 PRELIMINARY NOTIONS

We introduce a number of concepts and pieces of notation as some necessary but, we hope, simple preliminaries. The following comprises, we believe, the total notation of any special kind needed for assimilating the remainder of this book.

(i) $=, \neq$ equality, inequality

These symbols are certainly familiar in arithmetic, expressing the relationship of equality and inequality between two numerical quantities. We shall use them freely to express equality and inequality between any quantities, not necessarily numeric ones. Other types of quantities (sets, logical values, for example) will be introduced and defined.

(ii) $\triangleq$ is defined

This is used to define the notation or symbol to the left; the definition appears to the right. The $\triangleq$ may be read as 'equals by definition'.

(iii) $\mathbb{N}$ Natural Numbers

This denotes the 'natural numbers' $1, 2, 3, 4, \ldots$.

(iv) $\mathbb{N}_0$ Natural Numbers $+0$

The numbers $0, 1, 2, \ldots$.

(v) $>, \geqslant, <, \leqslant$ Inequalities

These symbols denote the familiar relations of greater than, greater than or equal to, less than, less than or equal to, between numeric quantities. They are not assumed to have any meaning for non-numeric quantities; for example, $7 \leqslant 8$, $9 \geqslant 9$.

(vi) . $\mathbb{Z}$ Integers

The integers ..., $-3, -2, -1, 0, 1, 2, 3, \ldots$.

(vii) $\mathbb{Q}$ Rationals

These are the numeric 'fractions' formed by dividing one integer by another non-zero integer. For example

   1/2, 12/13

(viii) $\mathbb{R}$ Reals

The real numbers include the rational numbers and others such as $2^{\frac{1}{2}}$, e, etc. The concept of real numbers is assumed to be familiar to readers.

(ix)   mod   modulus

The modulus of x, written mod x, or sometimes $|x|$, is that number which is the 'absolute value' of x; that is, if $x \geqslant 0$, then mod $x = x$, and if $x < 0$ then mod $x = -x$.

(x) $\langle \ldots \rangle$ sequences, lists

The idea of an ordered sequence or list is assumed as it is a familiar every-day object (individuals in a bus queue, a list of names on a piece of paper), and is used in examples in the early chapters. It is defined more rigorously in chapter 5. The expression $\langle 1, 2, 6, 4 \rangle$ means the sequence of numbers 1, 2, 6, 4 in that order.

(xi)   Finite, infinite quantities

If a quantity can be expressed as a number (especially a natural number) we say that the quantity is finite. For example, the number of integers between -1000 and $+1000$ inclusive (2001) or the population of the world at a specific moment, are finite. On the other hand the number of points on the real line between -1000 and $+1000$ or the number of rational numbers whose value v is such that $0 < v < 1$ are infinite (that is, not finite).

(xii)   The distinction between objects and their names (denotations)

We normally make this distinction be enclosing the named object in quotation marks. In most programming languages one denotes the value of an identifier by writing its name V say, whereas one denotes the name of the identifier by writing its name in quotes "V". Here "V" in a programming language normally denotes (stands for) the character string. We follow and generalise the application of this convention. For example, numbers are abstract entities; the number 4 is an abstract notion which has a variety of possible names: "four", "4" and "$2 + 2$" are three different ways of denoting 4.

1.4 THEOREMS AND PROOFS

Axioms, Theorems and Observations

Any mathematical system or topic, such as arithmetic or Euclidean geometry, has a number of *axioms* which form its basis, in the sense that these axioms are given as facts and other facts in the system can be deduced from them. These other facts deducible from the axioms are called *theorems*. Thus in principle any fact, no matter how trivial, which can be deduced from the axioms, is a theorem; this includes the axioms themselves, since one can trivially deduce an axiom from itself. A theorem does not have to be a significant step in the development of the topic. However convention usually classifies as 'theorems' only those inferred facts which are important in some way, and so we shall generally refer to the less important theorems, included for pedagogy or demonstration, as *observations*.

As an example of an axiomatic system, in the last century G. Peano produced a set of axioms for the natural numbers. With these axioms "$2+2=4$" is a theorem deducible from the definition of "$+$" and the axioms.

Proofs

Proofs play a crucial role in the disciplined engineering of computer programs. The behaviour and properties of a program can be determined before it is executed by means of mathematical proof. We shall give here an outline only of the techniques of proof by *natural deduction*. This is given a formal treatment in other works: see for example Tennant (1978), or Quine (1980) for a treatment which relates natural logic to natural language forms.

The proof method known as 'natural deduction', which, in general, we follow in this book, can be given as a set of formal rules of symbol manipulation. This is done in, for example, Tennant (1978). However these rules are a formalisation of rules which fall into a natural language framework: hence the phrase 'natural deduction'. Quine (1980) shows many connections between natural language forms and the formal rules.

Proofs in the natural deduction style are in general of two main kinds: by deduction and by contradiction. Proofs by deduction consist of the application of a number of steps, each consisting of a set of premises and a conclusion. The premises are statements which, at the particular stage in the argument, are taken to be true. The conclusion (or *inference*) is a statement which can be deduced from the premises by the application either of a rule (which will be a 'natural deduction' rule) or of a more general result which has been established.

For example, one rule is that if we know that statements A and B are true, we can deduce the truth of the composite statement A AND B. Another such rule is that if we know that the statement $P(X)$ is true where $P(X)$ is a statement containing the (free) variable X and X is arbitrarily chosen from a set of objects S, then we can deduce that for all objects X chosen from S, $P(X)$ is true.

There are a lot of rules which appear to be very laboured when written down in this way. They mostly appeal to intuition, and proofs in a natural deduction style as a result appear as linguistic arguments.

Proofs by deduction thus take the form of a chain of inferences, starting from one or more given premises and reaching the desired conclusion. Each step or inference proceeds from previously established results in the chain by the application of a rule. Often steps will be omitted, where it is supposed that it is clear that the detailed argument can easily be supplied.

The second kind of proof is a proof by contradiction. The method is as follows. Suppose that we wish to show that the conclusion C can be deduced from the premises $P1, \ldots, Pn$. We suppose that $P1, \ldots, Pn$ are true and that NOT C is true. If from these $(n + 1)$ premises we can deduce a contradiction, that is that some statement Q is both true and not true, then we can deduce that C can be deduced from $P1, \ldots, Pn$.

Let us take an example. There are two premises

(i) If it rains I shall eat my hat.
(ii) I shall never eat my hat.

The conclusion to be proved is

"It will never rain."

We show two proofs, the first by deduction.

If it rains I shall eat my hat.   (premise)
Therefore: If I do not eat my hat it is not raining.
I shall never eat my hat.   (premise)
Therefore: It will never rain.

The second proof is by contradiction. Suppose it is raining.

If it rains I shall eat my hat.   (premise)
Therefore: I shall eat my hat.
But: I shall never eat my hat.   (premise)
This is a contradiction. Therefore the supposition "It is raining" is always false—that is, it will never rain.

A proof may consist of a number of large steps, each of which is proved by a variety of methods. Thus a proof may comprise a mixture of sub-proofs, some by deduction and some by contradiction.

A proof may be achieved by a strategy of sub-goals. Here the intermediate results are stated to start with, and expressed as goals: "if we can prove P, Q, R, then we shall be able to prove B". Then it is shown how B can be proved from P, Q, R. Finally the truth of the intermediate results is demonstrated provable from the initial premises. When taken to its extreme form a proof by sub-goals reads as a deductive proof written in reverse. Suppose we have to prove R from premise P. We first propose the sub-goal $G1$ and show that R is easily deducible from $G1$. The problem then resolves to proving $G1$ from P. We then propose the sub-goal $G2$ and show that $G1$ is deducible from $G2$. The problem then resolves to proving $G2$ from P. This process continues until the problem resolves to proving Gn from P and then showing that Gn is immediately deducible from P.

The example above proved in this fashion is as follows. We have to prove "It will never rain" from the premises "I shall never eat my hat" and "If it rains I shall eat my hat". We take as a sub-goal G the statement "If I do not eat my hat it is not raining". Then

> I shall never eat my hat.   (premise)
> If I do not eat my hat it is not raining.   (G)
> Therefore it will never rain.

It remains to prove G.

> If it rains I shall eat my hat.   (premise)
> Therefore if I do not eat my hat it is not raining.

This type of proof is not really any different from a proof by deduction; it is merely presented in a different order. However, variations in presentation can make a large difference to the ease of understanding a proof. It is in general advisable to present a proof in a way which suggests the thought processes by which one conceived it.

The last proof method we shall cover here is a proof by induction. Proof by induction can be used if one is to prove the truth of some property with respect to all members of a set whose members can be constructed by a finite number of steps from a finite number of initial members. The choice of available initial members and the choice of available steps must be finite. The usual example of such a set is the set of natural numbers 1, 2, 3, ... "1" is the initial member. We can construct any other natural number by adding 1 to (or 'taking the successor of') 1 a finite number of times. For example, the number 97 can be constructed by adding 1 96 times to 1.

The structure of a proof by induction is as follows. Suppose one wishes to prove property $P(n)$ is true of all members n of the set in question. The induction principle allows that if one has shown that (a) P is true of all the initial members of the set and that (b) if $P(n)$ is true then P is true of all members of the set which can be constructed from n in one step, then one may deduce that P is true of all members of the set.

This is more simply put if we keep to natural numbers as the set in question. The induction principle then states that if one proves $P(1)$ and that $P(n+1)$ is deducible from $P(n)$, then one may deduce that $P(n)$ is true for any $n \geqslant 1$. Consider the following example. We wish to prove the sum of all numbers 1 to n is

$$n(n+1)/2$$

We first show that it is true for the sum of all numbers 1 to 1

$$1(1+1)/2 = 1$$

We then have to show that if the sum S_n of all numbers 1 to n is

$$n(n+1)/2$$

then the sum S_{n+1} is

$$(n+1)(n+2)/2$$

Proof

$$
\begin{aligned}
S_{n+1} &= S_n + n + 1 \\
&= n(n+1)/2 + n + 1 \\
&= (n(n+1) + 2(n+1))/2 \\
&= (n+2)(n+1)/2 \\
&= (n+1)(n+2)/2
\end{aligned}
$$

The result, namely that

$$S_n = n(n+1)/2$$

follows by the induction principle.

We have mentioned sets in this explanation of induction. Sets are dealt with more fully in chapter 2. If readers find the reference to sets confusing here, they are advised to consider the explanation of induction simply in terms of natural numbers.

Notation

The end of a proof is usually marked by a box symbol, □. In more traditional works the letters QED (*quod erat demonstrandum*) are used. In this book we shall normally use the box symbol.

1.5 STRUCTURE OF THIS BOOK

Each chapter is preceded by a summary of the notation introduced, and followed by a summary of the results. These are intended to provide a ready reference to both definitions of new concepts and to theorems and other results. Chapters 2, 3 and 4 concern the most elementary notions of set theory and logic. They may be skimmed lightly by readers who are already familiar with some basic discrete mathematics, but nonetheless these readers are urged to consider the examples and workshop exercises. The reason is that the *application* of discrete mathematics to software engineering problems can be unfamiliar even to accomplished mathematicians. Also the examples relating to software engineering are thematic and distributed liberally throughout the book. These are all in the content of topics typically the subject of treatment by computer programs. They are deliberately continued from one chapter to the next and developed in the light of the further material which has been studied in order to demonstrate its application. A plentitude of these thematic examples is provided in order that, given the great variety of readers' experiences, at least one will strike the necessary cognitive chord in each reader and communicate the connection between the mathematical ideas and their application in software engineering.

The workshop exercises are likewise thematic. They start at the end of chapter 2 and require the application of each new idea as it is encountered. An introduction to the workshop exercises is given at the end of this chapter.

The workshop exercises are intended for group or class discussion and solution, under the tuition of an experienced supervisor. They constitute substantial projects and will, one hopes, encourage and convince readers of the application of these mathematical fundamentals to real projects.

There are other exercises throughout the book. These are for individual solution and are designed to enhance the skill of the reader. It is important to do these. Mathematics involves a skill in manipulating symbols, as well as an understanding of concepts. To develop the skill requires practice in the same way as learning a foreign language or a musical instrument (although I do not think it is nearly as hard). Answers to these exercises are given at the end of the book.

1.6 EXERCISE 1.1 (Workshop)

Take a recent programming task you have had to do, and write as precisely as you can a statement in natural language (assisted if you wish by diagrams etc.) of what it is that program is required to do. Avoid as far as possible any statement or prejudice as to how the program is to work, the nature of its internal data, etc.

PART I: Foundations

All computer programs manipulate information. We may be accustomed to thinking of all this information as being numeric, but in fact much of the time we are merely using numbers to represent information of some other kind. Programs ultimately interface with the real world. Data in the program represents, at some stage, real world information, whether this consists of information read from physical sensors in an embedded software system, information about employees in a personnel file (their names, addresses, ages, salaries), the positions of pieces on a chess board for a chess-playing program, or the relationships between entities in a school time-table (the names of the teachers, the classes, the lessons, the periods in which they take place).

To discuss and define this information in abstract terms, that is, without considering the way it is represented (its implementation) we need a suitable abstract language. Furthermore this abstract language must be defined precisely enough that we can reason about it. We need to be able to distinguish computer data which correctly represents the information from that which does not. We need to be able to draw deductions from the information, to infer its properties, and to determine whether a program which manipulates data representing the information is correct.

The language of Mathematics provides us with this flexibility, abstraction and precision. Set theory forms the basis of many topics in mathematics and gives us a simple and powerful way of defining information of practically any kind. Set theory is taught in schools, but its application to problem-solving is less well emphasised in the usual curricula. Information is what Software Engineering is all about, and since set theory gives us the language for defining it, we embark in chapter 2 on a simple exposition of sets and many examples illustrating their use in defining the information which computer programs typically process.

The second mathematical topic we require in order to reason about the information of our problems is Logic. This is introduced in chapter 3, so that after those two chapters we have the basic equipment to begin to describe in abstract terms the problems which our programs are required to solve.

The language of logic introduced in chapter 3 enables us to define relationships between sets and to construct more elaborate kinds of sets. This development is done in chapter 4, Sets and their Operations, so that by the end of Part I the reader has completed an apprenticeship and is able to provide abstract problem definitions with some facility.

Chapter 2 Sets

NOTATION INTRODUCED

Notation	Concept	Reference
$=$	equality of sets	Section 2.2
$\in$	'is a member of' or 'belongs to'	Section 2.2
$\notin$	negation of $\in$	Section 2.2
$\{\ldots\}$	set construction delimiters	Section 2.2

Reading instructions. While many readers have met sets in mathematics, they may be less accustomed to applying set theoretic ideas to real-world problems or computer applications. Those already familiar with set theory may scan this chapter quickly, but should take note of the examples.

2.1 INTRODUCTION

Definition 2.1

A *set* is a collection of distinct objects of any kind. The distinct objects in a set are called the *members* or *elements* of the set.

This is a highly informal definition but is workable for our purposes. Later we shall refine and restrict it in order to avoid certain technical problems. However we can learn how to use sets, manipulate them and apply them to real-life problems without a more precise definition, just as we learned to count, add up and do many other quite sophisticated operations with numbers without first grasping a formal definition of them.

Example 2.1

Figure 2.1 shows an example of a set which contains five members. To illustrate the principle that a set may contain any objects, the members chosen for this set are very different from each other. Some are very

Figure 2.1

concrete (my house), others are more abstract (the number 34) and a further member is a name of another member ("34" is the name of 34; see section 1.3 of chapter 1). The illustration in figure 2.1 denotes a set; that is to say, the closed line is by convention taken to enclose denotations of the objects which are members of the set. The phrase 'farmer Giles's pig' should be taken to denote a unique object, namely a particular identifiable pig. Likewise for the phrase 'my house', 'the Bible', etc. The sequence of characters '34' in the bottom left of the figure denotes a unique abstract object, namely the number 34, and that in the top right-hand corner, ' "34" ', denotes the unique character sequence "34".

The important point, perhaps laboured above, is that the members of a set must all be unique identifiable objects.

Thus we can have a set containing a member of one kind and another member which is the name of the first kind of member. This facility is particularly useful in list-processing applications, where we find 'records', which are a computer representation of sets, containing a pointer or reference to another record.

Notation

A set can be represented by listing its elements between braces ({ and }). We may also give names to sets, thus

Let $S \triangleq$ {farmer Giles's pig, "34", my house, 34, the Bible}

A set is identified entirely by its distinct members, regardless of order or repetition, so that all the following are different ways of identifying the same set S.

Example 2.2

{farmer Giles's pig, "34", my house, my house, 34, the Bible}
{the Bible, 34, "34", farmer Giles's pig, my house}
{the Bible, 34, farmer Giles's pig, 34, 34, "34", my house}

In other words, a set is defined by its members, not by the way in which it is defined. One conventionally tends not to duplicate members of a set in this way, but if repetitions do occur, they do not affect the membership, and hence the definition, of the set.

Defining Sets: Explicit Notation

We can define a set by simply listing its members, as we have already seen: the notation is to separate the members by commas and enclose them in { } braces

S = {farmer Giles's pig, "34", my house, 34, the Bible}
T = {34, {farmer Giles's pig, "34", my house, 34, the Bible}}
       = {34, S}

Example 2.3

Sets may contain many members. Figure 2.2 shows a set with an infinite number of members.

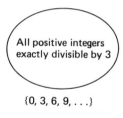

All positive integers
exactly divisible by 3

{0, 3, 6, 9, . . .}

Figure 2.2

Example 2.4

Since sets may contain as members objects of any kind, the members of a set may themselves be sets. Remembering the set we called 'S' in example 2.1, figure 2.3 shows a new set. The set in figure 2.3 has two members.

34   S

Figure 2.3

One is the number 34, the other is another set. This other set has five members. Thus if we call our new set 'T'

Let T be {34, S} .

Another way of defining T would be

Let T be $\{34, \{farmer Giles's pig, "34", my house, 34, the Bible\}\}$

or as illustrated in pictures in figure 2.4. One may observe a

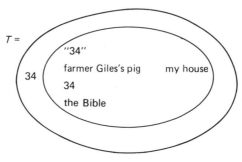

Figure 2.4

correspondence between the pictorial depiction of a set as in figure 2.4 and its explicit definition. Each closed line delimiting the members of a set corresponds to a pair of braces.

It should be clear from these illustrations that T has two members, even though one of its members is a set S with five members. The members of S are not automatically members of T. Thus T is a different set from the set U defined

Let $U = \{S\} = \{\{farmer Giles's pig, "34", my house, 34, the Bible\}\}$

or in pictures in figure 2.4. U is a set with one member only. 34 is not a member of U, but is a member of T.

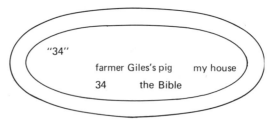

Figure 2.5

Exercise 2.1

Write expressions for the following sets

(a) The set of natural numbers from 1 to 10 which are multiples of 2.
(b) The set of letters a, b, c.

2.2 EXAMPLES AND NOTATION

Example 2.5: Car Supply Database

Sets of various kinds of objects occur in all manner of problems and applications. A stock control or ordering system for a commercial product, such as a motor car, will have intrinsic to it several sets of alternative variants of the product. The variations of a particular brand and model of a car may be represented by sets such as

$B \triangleq \{\text{estate, saloon, sports, \ldots}\}$—the various body styles

$C \triangleq \{\ldots\}$—available colours

$Eng \triangleq \{\ldots\}$—different engine sizes

$T \triangleq \{\ldots\}$—interior trim styles

Example 2.6: Bus Routes

If we were trying to write a program to schedule buses along a number of routes, we might start off with the following sets

$R \triangleq \{\ldots\}$—all the different routes

$T \triangleq \{\ldots\}$—the different types of buses

and for each type t of bus—that is, for each member t of T

$B_t = \{\ldots\}$—the individual buses of type t

Note our freedom of use of notation: for names of things, including sets, we have used single letters (B, T), strings of letters (Eng), or subscripted names (B_t) just as if we were dealing with any other mathematical objects such as real numbers. Also, as in other mathematics, we feel free to use the same name to represent different things in different contexts provided no confusion is likely to arise. (The rules of programming languages are much more pedantic in this respect. Scope rules define exactly whether two occurrences of a name may or may not denote the same object.)

Notation

Equality

We have said that a set is identified by its members. In other words, if two sets P and G have exactly the same members, they are identical. We can write $P = Q$.

Exercise 2.2

Which pairs of the following sets are equal

$$A = \{0, 2, 4, 6\}$$
$$B = \{4, 2, 0, 6\}$$
$$C = \{1, 2, 3, 1\}$$
$$D = \{2, 2, 3, 3, 1\}$$
$$E = \{0, 1, 2\}$$
$$F = \{1, 1, 2\}$$

Membership

If a specific object a is a member of a set S, we write

$$a \in S$$

The '$\in$' symbol, the Greek letter epsilon, is the set membership operator.
'$\notin$' is the negation of membership. The convention of putting a /
through the relation follows the usual practice as with $=$ (equals) and $\neq$
(not equals) for example.

Example 2.7

In our earlier sets S and T

$$34 \in S$$
$$34 \in T$$
$$S \in T$$
$$\text{"34"} \in S$$

but

$$\text{"34"} \notin T$$

The above relations may be read

$34 \in S$	'Thirty-four is a member of S'
$34 \in T$	'Thirty-four is a member of T'

but

$\text{"34"} \notin T$	' "Thirty-four" is not a member of T'

or one may read 'belongs to' in place of 'is a member of'.
  Thus if S is a set, expressions of the form

$$x \in S$$
$$y \in S$$

are either True or False.

Exercise 2.3

Let $S = \{a, b, c\}$. Which of the following statements is true?

(a) $a \in S$
(b) $\{a\} \in S$
(c) $\{b\} \notin S$
(d) $\{a, b\} \in S$
(e) $d \notin S$

Example 2.8: Bus Routes

In a certain (fictitious!) town, there are a set of districts and also a set of bus routes. Each bus route consists of a sequence of districts. A sequence is not a set because the order of items in the sequence is relevant, and the same item may occur more than once in the sequence. Readers will be familiar with sequences of numbers, and indeed of many kinds of objects. In chapter 5 we shall provide a more rigorous definition of sequences, but for the purposes of these examples the intuitive notion of a list or sequence will serve us. The sequences are denoted by the names of the districts separated by commas and enclosed in special brackets $\langle \ \rangle$. For example

$\langle$Winchmore Hill, Palmers Green, Wood Green, Seven Sisters$\rangle$

The set of districts is as follows

$D = \{$Winchmore Hill, Palmers Green, Bowes Park, Wood Green, Turnpike Lane, Seven Sisters, Enfield, Manor House, Stamford Hill, Stoke Newington, Finsbury Park, Lower Edmonton, Pickets Lock, ...$\}$

The bus routes can be represented as follows

$Busr = \{\langle$Winchmore Hill, Palmers Green, Wood Green, Seven Sisters$\rangle$,
$\langle$Enfield, Lower Edmonton, Pickets Lock$\rangle$,
$\langle$Wood Green, Turnpike Lane, Seven Sisters, Stamford Hill, Stoke Newington$\rangle$ ...
...
...
...
...
...$\}$

Thus *Busr* is a set of sequences. Then

$\langle$Enfield, Lower Edmonton, Pickets Lock$\rangle \in Busr$

whereas (probably)

⟨Finsbury Park, Enfield, Stoke Newington, Bowes Park⟩ ∉ *Busr*

(As this fictitious town is rather large, not all the routes or districts are listed!)

Members of sets may themselves be sets, or they may not, as in the case of districts above. We make the distinction between objects which are sets and objects which are not.

Definition 2.2

An *Atom* is an object which is a member of a set and is not itself a set.

Example 2.9: Part Numbers

A product, which has an identification number 347 622, consists of a number of parts

347 622 = {266 982, 37 622A, 4011B, 361 044, 2377A}

Parts 361 044, 4011B, and 2377A are atomic; parts 37 622A and 266 982 are sub-assemblies composed of further components as follows

266 982 = {4011B, 3186, 7922C}
37 622A = {3186, 2922C, 4099, 4011B}

Clearly, for a concrete practical application these sets would have to have certain properties: no set should be a member of itself; one should not be able to find a sequence of sets $S_1 \in S_2 \in S_3 \ldots \in S_n$ with $S_n \in S_1$; the number of sets in the system must be finite, etc. As we develop these ideas further, we shall discover how to express these kinds of constraints in a concise and rigorous way.

Defining Sets: Implicit Notation

Another way of defining a set is to say that it consists of all objects with certain properties. That is, one asserts that each member of the set fulfils some condition. The condition is expressed by an expression involving a variable, the value of the expression being either True or False. Such an expression is called a Predicate. We shall define this more rigorously in chapter 3.

$\{x | x \in A \text{ and } P(x)\}$

means the set of all members x of A such that $P(x)$ is True.

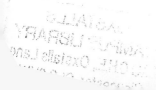

Example 2.10

Here is an example where the predicate is expressed in words

$\{x|x \text{ is a positive integer divisible by } 3\}$

that is

$\{x|x \in \mathbb{N}_0 \text{ and } x \bmod 3 = 0\}$

Another example:

$\{x|x \text{ is a positive integer less than } 7\}$

that is

$\{x|x \in \mathbb{N}_0 \text{ and } x < 7\}$
$= \{0, 1, 2, 3, 4, 5, 6\}$

Exercise 2.4

Provide an explicit definition of the following sets

(a) $\{x|x \in \mathbb{Z} \text{ and } x^2 < 10\}$ (recall that $\mathbb{Z}$ is the set of integers).
(b) $\{n|n \in \mathbb{N} \text{ and } 5 = 2 + n\}$.

Write in formal notation

(c) The set of all multiples of 5.
(d) The numbers 1000 to 2000.

Example 2.11: Character Sets

In a computer system there is a character set, which consists of a set of characters

$$C = \{\text{"A", "B", \ldots, "Z", "a", \ldots, "z", " ", ".", ",", ";", \ldots, "0", "1",} \\ \text{\ldots, "9"}\}$$

Associated with each member of the character set there is an integer $N(c)$. the numerical value of character c. Some useful subsets of this character set are as follows

$Digitch = \{x|x \in C \text{ and } N(x) \geqslant N(\text{"0"}) \text{ and } N(x) \leqslant N(\text{"9"})\}$
$Alphanumeric = \{x|x \in C \text{ and } (N(x) \geqslant N(\text{"A"}) \text{ and } N(x) \leqslant N(\text{"Z"})) \text{ or} \\ (N(x) \geqslant (\text{"a"}) \text{ and } N(x) \leqslant N(\text{"z"})) \text{ or} \\ (N(x) \geqslant N(\text{"0"}) \text{ and } N(x) \leqslant N(\text{"9"}))\}$

Note that we have slyly slipped some function notation into this example. Readers unfamiliar with this can pass on to the next section.

2.3 MORE EXAMPLES

Example 2.12: Personnel File

A personnel file contains the names of all the employees in a firm, their ages, grades and salaries. The information is therefore drawn from the following sets

  Names, which are sequences of *Name-characters*

where

$$Name\text{-}characters = \{x | x \in C \text{ and } (x \in Alphabetic \text{ or } x = \text{"."} \text{ or } x = \text{"-"})\}$$

and

$$Alphabetic = \{x | x \in C \text{ and } ((N(x) \geqslant N(\text{"A"}) \text{ and } N(x) \leqslant N(\text{"Z"})) \text{ or } (N(x) \geqslant N(\text{"a"}) \text{ and } N(x) \leqslant N(\text{"z"})))\}$$

C and N being defined in example 2.11

$$Ages = \{n | n \in \mathbb{N}_0 \text{ and } n \geqslant 16 \text{ and } n \leqslant 65\}$$

where $\mathbb{N}_0$ is the set of natural numbers together with zero.

  Grades = {Technical Assistant 1, Technical Assistant 2, Technical Officer 1, Technical Officer 2, Senior Technical Officer, Principal Technical Officer, Senior Principal Technical Officer, Chief Technical Officer, Section Leader, Department Manager, Division Manager, Director}

  $Salaries = \{n | n \in \mathbb{N}_0 \text{ and } n < 1\,000\,000\}$

(This set is not inflation-proof!)

Example 2.13: File-processing

A file contains data which reflects some real-world state of affairs, such as the numbers of different items in a stock, the status of modules in a very large programming system which is under development, the details of the bank accounts of a number of bank customers, or the experiments and results which have been carried out on individuals in a selected population in a biological/sociological research programme. The subject matter of the contents of the file is irrelevant: the file consists of a number of records and may have a number of operations or transactions performed upon it.

  There are various sets involved in this system. The file itself is a member of the set F of all sequences of records, where the records themselves will be members of some set R. We shall see in chapter 5 how to define 'sequence' formally.

The transactions will form a set, for example

$$Transactions = \{Add, Delete, Update, Purge\}$$

Associated with each transaction will be an ordered sequence of parameters each belonging to a set, for example

$Add:—\langle record \rangle$
$Update:—\langle record, info \rangle$

where

$record \in \mathbb{R}$
$info \in INF$

where INF is a set of relevant information such as readings

$$Readings = \{n | n \in \mathbb{R} \text{ and } n > 10^{-6} \text{ and } n < 10^{6}\}$$

Example 2.14 (from Artificial Intelligence): a Chess Program

If we were contemplating writing a chess-playing program, we might start to define the problem by considering the following sets

A set M of piece types, $= \{$Pawn, Rook, Knight, Bishop, King, Queen$\}$
A set C of Colours $= \{$Black, White$\}$
A set F of files $= \{A, B, C, D, E, F, G, H\}$
A set R of rows $= \{1, 2, 3, 4, 5, 6, 7, 8\}$

For each member c of C, a position Pos_c is a set of pieces, where each member of Pos_c is associated with

a row in R
a file in F
a piece type in M

Associated with each piece type m in M there is a set of rules (for possible moves). How to express these rules may be a difficult problem!

Part of our chess-playing program will probably be a procedure which is called with a parameter belonging to the set C (colour) and delivers as result a set of all possible moves. This procedure applies the set of rules associated with the piece type which in turn is associated with each piece in Pos_c, where c is the value of the input parameter, to the total position on the board (which is given by Pos_{Black} and Pos_{White}), in order to produce all the possible moves for colour c.

2.4 WORKSHOP EXERCISES

Exercise 2.5 (Workshop)

Consider carefully the statement of the problem or problems which you provided at the end of chapter 1. In each case identify and name sets to which all items of data in the problem(s) belong.

Exercise 2.6 (Workshop)

In a nuclear power station, there is a reactor and a number of sensors which sense conditions of the reactor—its temperature at different positions and energy output—and conditions of the coolant—temperature and pressure at various places and rate of flow through a number of paths through the reactor. All these conditions are each sensed by several sensors and their readings are monitored by a computer program. The program also records its own state in a selective fashion, and may thus react to its own history. The program outputs information to a collection of servo-mechanisms which control the positions of rods inserted into the reactive material in the reactor: these control the activity of the reactor. The program also outputs readings to dials, lights and VDU screens, and will also indicate alarm situations on lamps (red/green) when readings from the sensors or combinations of them go out of range.

The program may also react to input requests to display any of the information available to it on an individual VDU or several selected VDUs. However, alarm information must not be obscured by such requests.

Define the sets which comprise the data of this system, both input to, intrinsic to and output from the program.

Exercise 2.7 (Workshop)

You wish to program your personal computer to provide you and your family with a general diary/reminder system. First jot down in English what you would like this program to do, and then define the sets to which all the data associated with the program belong. Does this give you any further ideas about what the program ought to do? If so, amend the English language specification, redefine the sets of the data, and continue iterating until you are satisfied that you have a useful system which is feasible to implement on a personal computer. Reflect on whether and to what extent the formalisation of the data into sets has made you reconsider your original requirements!

Exercise 2.8 (*Workshop*)

Those who are anxious for further exercises may cycle through the same English specification—data set definition for the following problems

(i)   one aspect of a space-borne computer system on a space-shuttle.
(ii) a fault-tolerant replicated real-time control system.

Chapter 3 Logic

NOTATION INTRODUCED

Notation	Concept	Reference
~	not	Def. 3.1
∧	and	Def. 3.2
∨	or	Def. 3.3
⇒	implies	Def. 3.4
≡	equivalence	Def. 3.5
∀	for all	Def. 3.11
∃	there exists	Def. 3.12

Reading instructions. This chapter concentrates on a simple form of predicate logic, easily assimilated and sufficient for the purpose of constructing predicates for defining sets implicitly and expressing constraints on set members and their interrelationships. Those to whom this treatment is familiar may scan this chapter rapidly, taking note of the application of the ideas illustrated in the examples. The exercises will help to show how the principles can be used to provide more precise specifications.

3.1 INTRODUCTION

Treatments

Logic can be treated at various levels of rigour. At this stage we approach the topic in a less rigorous manner which will probably appear familiar on account of its similarity to arithmetic and algebra.

Logic as an Algebra

We have all learned 'Algebra' at school; what then is 'an algebra'? Any simple algebraic system consists of a set of values and a finite set of operations. In school algebra, the set of values is the real numbers, the operations $+$, $-$, $/$, $*$. Later we learn further operations such as exponentiation and logarithms. An operation takes one or more argument values from the value space and produces a new value as result. In fact '$-$' represents two operations, subtraction which takes two arguments and negation which takes only one argument. In the algebra of natural numbers the set of values is the numbers $\mathbb{N}$, and the operations are Successor and Addition. In integer arithmetic the value space consists of $\mathbb{Z}$, the positive and negative integers with zero, and the operations are Successor, Addition, Subtraction and Multiplication

In logic, the set of values is {True, False} and the operations are Not, And, Or, Implies and Equivalence. 'Not' is a monadic operation—that is, it takes one argument—and the others are dyadic and take two arguments. These values are called Boolean values after the British mathematician George Boole who wrote a mathematical treatment of propositions and their combination, called *The Laws of Thought* in the nineteenth century. The five operations are represented by symbols called Boolean operators

$\sim$	not
$\wedge$	and
$\vee$	or
$\Rightarrow$	implies
$\Leftrightarrow$	or $\equiv$ equivalence

Other symbols are also often used. For example

$\neg$	for not
$\rightarrow$, $\supset$	for implies

This should appear familiar to users of high level languages such as Pascal, Algol, Ada and Chill.

As in other algebraic systems, in addition to constant values, we may use variables which may take values from the set of values in the algebra, and we can construct expressions using variable names, operators and constants. In many texts, in place of the symbols, the names of the operators are written in bold type or underlined.

Notation

We shall refer to the Algebra {True, False} with the operations mentioned above as '$\mathbb{B}$'.

Here are some examples of expressions in $\mathbb{B}$

$p \wedge \sim q$

$(p \vee r) \wedge \sim (q \vee s)$

For any algebra, the meanings of the operations can be defined by means of a collection of fundamental rules called axioms. In the late nineteenth century the mathematician G. Peano devised a set of axioms which defined the system of natural numbers. However, nobody learns Peano's axioms before they can count, and historically numbers of all kinds were widely used and applied before Peano produced his axiomatisation of numbers. Likewise Boolean algebra can be defined axiomatically (and actually many different instances of algebras, of which our familiar two-valued kind is but one, will satisfy the axioms), but we shall not introduce the axiomatic treatment at this stage.

3.2 LOGICAL OPERATIONS

Because our Boolean algebra $\mathbb{B}$ has a finite underlying set, the operators can be totally defined by finite tables. The familiar multiplication tables which children learn (or used to learn) at school defined the meaning of $*$ applied to a finite subset of the integers. To define $*$ totally we need some other technique such as a rule in the form of an equation or an algorithm based on such a rule.

However, for the operations in $\mathbb{B}$ we can manage with finite tables, the so-called truth tables, as follows (we use T, F interchangeably to denote True, False).

Definition 3.1

Not

	x	
	T	F
$\sim x$	F	T

Definition 3.2

And

$x \wedge y$		T	F
	T	T	F
y			
	F	F	F

Definition 3.3

Or

		x	
		T	F
$x \lor y$	T	T	T
y			
	F	T	F

Definition 3.4

Implies

		x	
		T	F
$x \Rightarrow y$	T	T	T
y			
	F	F	T

Definition 3.5

Equivalence

		x	
		T	F
$x \equiv y$	T	T	F
y			
	F	F	T

Truth tables (familiar to digital electronic engineers) are useful for calculating the values of general Boolean expressions, and for establishing that two expressions are equivalent; that is, that they give the same value as each other for all combinations of the variables occurring in the expressions. As a special case, a truth table can also show that an expression has the value 'True' for all values of its variables. A Boolean expression whose value is always True is called a 'tautology'; one whose value is always False is a contradiction. Here are some examples.

Example 3.1

$x \Rightarrow y$ is the same as $\sim x \vee y$, that is

$$(x \Rightarrow y) \equiv (\sim x \vee y)$$

We construct the truth table as follows

a	b	c	d	e
x	y	$x \Rightarrow y$	$\sim x$	$\sim x \vee y$
T	T	T	F	T
T	F	F	F	F
F	T	T	T	T
F	F	T	T	T

Column c is constructed from columns a and b using the truth table for Implies. Column d is constructed from column a using the truth table for Not. Column e is constructed from columns d and b using the truth table for Or.

We can add a sixth column for the value of the total expression

f
$(x \Rightarrow y) \equiv (\sim x \vee y)$
T
T
T
T

and this column is constructed from columns c and e using the truth table for Equivalence.

Theorem 3.1: De Morgan's Laws

(1) $\sim (x \wedge y)$ and $\sim x \vee \sim y$ are equivalent.

Proof
The following truth table is derived from the tables for $\sim$, $\wedge$, $\vee$

x	y	$\sim x$	$\sim y$	$x \wedge y$	$\sim x \vee \sim y$	$\sim (x \wedge y)$
T	T	F	F	T	F	F
T	F	F	T	F	T	T
F	T	T	F	F	T	T
F	F	T	T	F	T	T

Hence $\sim(x \wedge y)$ and $\sim x \vee \sim y$ have the same values for all values of x and y. ☐

(2) $\sim(x \vee y)$ and $\sim x \wedge \sim y$ are equivalent.

Proof
Using the results above we can continue the truth table

x	y	$x \vee y$	$\sim(x \vee y)$	$\sim x \wedge \sim y$
T	T	T	F	F
T	F	T	F	F
F	T	T	F	F
F	F	F	T	T

and draw a similar conclusion. ☐

Properties of Operations

There are various important properties operations can have.

Definition 3.6
Given a general dyadic operation O, O is *Commutative* if

$$x O y = y O x$$

for all x, y in the underlying set.
 Thus in the algebras of integers and reals, $+$, $*$ are commutative but $-$ and $/$ are not.

Theorem 3.2

In $\mathbb{B}$ $\wedge$, $\vee$ and $\Leftrightarrow$ are commutative, but $\Rightarrow$ is not.

Small Exercise 3.1

Prove theorem 3.2 above using truth tables.

Definition 3.7
An operator O is *Associative* if

$$(x O y) O z \equiv z O (y O z)$$

for all x, y, z in the underlying set.
 Thus, again in the algebras of integers and reals, $+$ and $*$ are associative but again $-$ and $/$ are not.

Theorem 3.3

In $\mathbb{B}$, $\wedge$, $\vee$ and $\Leftrightarrow$ are associative.

Small Exercise 3.2

Prove theorem 3.3 above using truth tables. Is $\Rightarrow$ associative?

Definition 3.8
An operator O_1 is *Distributive* with respect to another operator O_2 if

$$x O_1 (y O_2 z) \equiv (x O_1 y) O_2 (x O_1 z)$$

for all x, y, z in the underlying set.
    In the algebras of integers and reals, $*$ is distributive with respect to $+$.

Theorem 3.4: Distributive Law

In Boolean algebra, $\wedge$ is distributive with respect to $\vee$ and $\vee$ is distributive with respect to $\wedge$; that is

$$x \vee (y \wedge z) = (x \vee y) \wedge (x \vee z)$$

and

$$x \wedge (y \vee z) = (x \wedge y) \vee (x \wedge z)$$

Small Exercise 3.3

Prove theorem 3.4 using truth tables.

More Small Exercises 3.4

Verify that the following pairs of expressions are equivalent (that is, have the same truth value)

(a) $\sim(\sim x)$, x
(b) $x \Leftrightarrow y$, $(x \Rightarrow y) \wedge (y \Rightarrow x)$

Rules (Theorems) and Expressions

Once we have established a rule—that is, a theorem—we can substitute any expression for a variable name in that rule, and obtain a new rule. For example, in integer arithmetic

$$x * (y + z) = (x * y) + (x * z)$$

We can derive an endless succession of further rules by substituting for x, *y*, or *z*

$$3*(y+z)=(3*y)+(3*z)$$
$$3*((a*7)+z)=(3*(a*7))+(3*z)$$

Likewise in $\mathbb{B}$, using the rule

$$\sim(\sim x)=x$$

we can derive

$$\sim(\sim\text{True})=\text{True}$$
$$\sim(\sim(x\Rightarrow\text{False}))=(x\Rightarrow\text{False})$$
$$\sim(\sim(a\Rightarrow b))=(a\Rightarrow b)$$

etc.

   Proofs of rules can be constructed by applying rules and substitutions in an appropriate sequence, starting from the initial rules (that is, the axioms). However, we shall not need to construct proofs at this stage.

   Brackets can become unwieldy. We can reduce their occurrence in two ways.

(i) If O is an associative operation, then

$$x\,O\,(y\,O\,z)$$

and

$$(x\,O\,y)\,O\,z$$

may be replaced by

$$x\,O\,y\,O\,z$$

where *x*, *y*, *z* are any sub-expressions. Hence above

$$(3*(a*7))$$

can be replaced by

$$(3*a*7)$$

(ii) We can establish priorities for operators, just as we do in arithmetic. The priorities for the arithmetic operators are

$$*\ /$$
$$+\ -$$

That is, multiplication and division are applied before addition and subtraction, so that

$$x+y*z$$

is taken to mean

$$x+(y*z)$$

In the same way the usual priorities for Boolean operations are defined as follows, in decreasing order

$\sim$

$\wedge$

$\vee$

$\Rightarrow$

$\equiv$

so that, for example, $\sim x \equiv \sim y$ means $(\sim x) \equiv (\sim y)$.

Thus in the above we could write

$3 * a * 7 + 3 * z$ and

$\sim \sim \text{True} \equiv \text{True}$

$\sim \sim (x \Rightarrow \text{False}) \equiv x \Rightarrow \text{False}$

$\sim \sim (a \Rightarrow b) \equiv a \Rightarrow b$

However, 'unnecessary' brackets should always be used if they are considered to improve readability.

3.3 PROPOSITIONS AND PREDICATES

Logic (that is, the algebra $\mathbb{B}$) has two applications which are relevant for our purposes. Firstly, it is an algebra of particular quantities (True and False) to be manipulated by the program in the same way as numbers and sets. It is likewise an algebra of quantities to be treated in the problem specification. Secondly, it is an algebra of quantities which stand for assertions about the values of which the problem treats. This means that computer programs are capable of a measure of self-reflexivity, in that they can contain assertions (standing as Boolean expressions) about values of their own data.

These two 'applications' are strongly related. A statement about a value in the problem can be considered as a new 'value' derived from an existing one. Let us first take some examples to motivate these ideas.

Example 3.2

Consider the following program which sets *max* to the maximum of two input values x and y.

Bool b;

$b := (x \geqslant y)$;

—assertion: $b = (x \geqslant y)$

if b **then** $max := x$ **else** $max := y$;

—assertion: $b \Rightarrow max = x$

—assertion: $\sim b \Rightarrow max = y$

Example 3.3: Personnel File

Suppose

$age \in Ages$

is the age of some individual. We could define

$ageband \triangleq age$ DIV 10

where DIV is an integer division operator giving an integer result, and ignoring any remainder, so that

$ageband * 10 \leqslant age < (ageband + 1) * 10$

Likewise we could define a Boolean value

$retiring\text{-}soon = age > 60$

'$age > 60$' is called a 'predicate'. Its truth value depends on the value of 'age'. If we instantiate 'age' with a value (such as 40) we obtain a proposition

$40 > 60 \Leftrightarrow \text{False}$

$63 > 60 \Leftrightarrow \text{True}$

etc.

Definition 3.9
A Proposition is a statement which has a value of True or False.

Definition 3.10
A Predicate is an expression containing variable symbols, such that when the variables are replaced by values, the expression is a proposition.
Examples of propositions are

$40 > 60$

$63 > 60$

Examples of predicates are

$age > 60$

$age > 60 \qquad age < 90$

$x \Rightarrow y \wedge p \Rightarrow q$

where x, y, p, q are Boolean variables, that is variables whose values are True or False.

George Boole originally proposed his algebra as a means of manipulating the truth values of propositions or statements. Thus, in place of variables in an expression, one may have a proposition

$$\text{The sun is shining} \equiv \text{The sun is shining}$$

$$\text{It snows in June} \Rightarrow \text{I'll eat my hat}$$

The second of these is a version of the colloquialism 'If...⟨some statement which you are sure is false⟩...I'll eat my hat (another statement you wish to be false!)'. This gives an illustration of the rule False⇒False which many people find counter-intuitive.

Other propositions can be provided by other branches of mathematics

$$x \leqslant y + z$$

$$a \in S$$

Arithmetic relations like $\leqslant$ and $=$ are examples of heterogeneous operators because the value of their result is of a different type from the value of their arguments.

Exercise 3.5

By setting variables to stand for propositions, write predicates which represent the following statements (the answer is supplied to the first one for guidance).

(a) If the sun is shining it is unlikely to rain.

Answer: let $P =$ the sun is shining
$Q =$ it is unlikely to snow

then $P \Rightarrow Q$.

(b) Fred's house is either green or red.
(c) London buses are painted green or red.
(d) It never rains but it pours.
(e) When on a cobbled road, my car will not go fast.
(f) Students take computer science or electronic engineering but not both.

Example 3.4: Car Supply Database

In our example of a stock control system for a motor car, there would probably be certain restrictions on the combinations of attributes which any one car may have.

Recall that

B is the set of body styles
C is the set of colours

E is the set of engine sizes

T is the set of interior trim styles

Let $b \in B$, $c \in C$, $e \in E$, $t \in T$ be the attributes of some car. Then we may have certain restrictions such as

$b = \text{sports} \Rightarrow e = 2 \text{ litres} \lor e = 3 \text{ litres}$

$c = \text{silver} \Rightarrow b = \text{sports}$

From these we can deduce various other properties, for example

$c = \text{silver} \Rightarrow e = 2 \text{ litres} \lor e = 3 \text{ litres}$

$e = 1 \text{ litre} \Rightarrow b \neq \text{sports}$

These properties are all examples of predicates.

Example 3.5: Bus Routes

In our example of bus routes, recall that

R is the set of routes

T is the set of types of bus

B_t where $t \in T$ is the set of buses of type t

It is likely that only buses of certain types will be used on each route, depending on the distance, likely number of passengers etc. We can therefore define the operator '*used-on*' such that, for each $r \in R$ and $t \in T$, t *used-on* r has a value of either True or False. Likewise, we can define an operator *alloc* which tells us which buses are allocated to a given route, such that for each $b_t \in B_t$ and $r \in B$, b_t *alloc* r is either True or False.

Since only buses of type t, where t *used-on* r, can be used on route r, we have a property of the system expressed (rather more concisely) by the predicates

$b_t \text{ } alloc \text{ } r \Rightarrow t \text{ } used\text{-}on \text{ } r$

Example 3.6: Character Sets

In our example of character sets, we considered the function N defined on members x of the character set C. Subsets of C can now be expressed in the notation we have just defined

$Digitch \triangleq \{x | x \in C \text{ and } N(x) \geq N(\text{``0''}) \land N(x) \leq N(\text{``9''})\}$

$Alphunumeric \triangleq \{x | x \in C \text{ and } N(x) \geq N(\text{``A''}) \land N(x) \leq N(\text{``Z''})$
$\lor N(x) \geq N(\text{``a''}) \land N(x) \leq N(\text{``z''})$
$\lor x \in Digitch\}$

Note that here we have reduced the numbers of brackets by using the priority of the $\land$ and $\lor$ operators.

In our Data Processing example we can similarly make use of the $\land$ and $\lor$ operators as follows.

$$Name\text{-}characters \triangleq \{x | x \in C \text{ and } x \in Alphabetic \lor x = \text{“.”}$$
$$\lor x = \text{“'”} \lor x = \text{“–”}\}$$

$$Alphabetic = \{x | x \in C \text{ and } N(x) \geqslant N(\text{“A”}) \land N \leqslant N(\text{“Z”})$$
$$\lor N(x) \geqslant N(\text{“a”}) \land N \leqslant (\text{“z”})\}$$

$$Ages = \{n | n \in \mathbb{N}_0 \text{ and } n \geqslant 16 \land n < 65\}$$

The combinations such as

$$x \in Alphabetic \lor x = \text{“.”} \lor x = \text{“'”} \lor x = \text{“–”}$$

are expressions: they are Boolean expressions and have Boolean values, and therefore they can be used wherever a Boolean value is required.

Because the expressions written here involve a free variable x and have a Boolean value, they are Predicates, and that is exactly what has to occur in an implicit set definition. Hence Boolean expressions are very useful for defining sets: indeed this is why we have introduced them at this stage. The operators in Boolean algebra enable us to construct elaborate predicates.

3.4 QUANTIFICATION

There is some further notational apparatus, called Quantification, which provides us with further power in defining Predicates.

Definition 3.11: The Universal Quantifier

$$\forall x \in X . P(x)$$

This is read as: 'For all x belonging to $X, P(x)$' where P is some predicate of x. The expression $\forall x \in X . P(x)$ has the value True or False. If $P(x)$ has no free variables other than x, then $\forall x . P(x)$ is itself a proposition. It has the value True if $P(x)$ is True for every value of x in the set X; otherwise it has the value False.

Example 3.7

The following propositions are true

$$\forall x \in \mathbb{N} . (x + 2) \in \mathbb{N}$$
$$\forall x \in \mathbb{N} . (x * 2) \in \mathbb{N}$$

The following propositions are false

$$\forall x \in \mathbb{N} . (x - 2) \in \mathbb{N}$$
$$\forall x \in \mathbb{N} . (x/2) \in \mathbb{N}$$

Definition 3.12: The Existential Quantifier

$\exists x \in X . P(x)$

This is read as: 'There exists a member x of X such that $P(x)$'. Again if $P(x)$ has no free variables other than x, $\exists x \in X . P(x)$ is a proposition, that is it has a value True or False. It has the value True if $P(x)$ is True for some value of x which belongs to the set X, otherwise it has the value False.

Example 3.8

The following propositions are true

$\exists x \in \mathbb{N} . 2 * x = 10$

$\exists x \in \mathbb{N} . (x - 2) = \mathbb{N}$

The following propositions are false

$\exists x \in \mathbb{N} . x < 0$

$\exists x \in \mathbb{N} . \forall y \in \mathbb{N} . x > y$

Notation

Other variations on notation abound in the literature, such as

$(\forall x \in X)(P(x))$

$\forall x \in X | P(x)$

For example

$(\forall x \in \mathbb{N})((x + 2) \in \mathbb{N})$

$\forall x \in \mathbb{N} | (x + 2) \in \mathbb{N}$

etc. Also, where the context makes it very clear what the set X is, this is sometimes omitted; for example

$\forall x . P(x)$

However, we shall always try to be explicit.
For a finite set $S = \{s_1, s_2, s_3, s_4, \ldots, s_n\}$,

$\forall x \in S . P(x) \equiv P(s_1) \wedge P(s_2) \ldots \wedge P(s_n)$

and

$\exists x \in S . P(x) \equiv P(s_1) \vee P(s_2) \ldots \vee P(s_n)$

Note that the priorities of the operators are very important here. '$\vee$' has higher priority than '$\equiv$'.

Example 3.9

We could define $m \operatorname{div} n$ (m is divisible by n) by

$$m \operatorname{div} n \triangleq (\exists x \in \mathbb{N})\ (m = n * x)$$

where $\mathbb{N}$ is the set of natural numbers. We may then define

$$Prime(m) \triangleq m \neq 1 \wedge \forall x \in \mathbb{N} . m \operatorname{div} x \Rightarrow x = m \vee x = 1$$

Theorem 3.5

For all predicates P

(i) $\forall x \in X . P(x) \equiv\ \sim \exists x \in X .\ \sim P(x)$
(ii) $\sim \forall x \in X .\ \sim P(x) \equiv \exists x \in X . P(x)$

Proof
$$\forall x \in X . P(x)$$

then for arbitrarily chosen x

$$P(x)$$

and hence for arbitrarily chosen x it is not the case that

$$\sim P(x)$$

that is, we cannot find x such that

$$\sim P(x)$$

therefore

$$\sim \exists x \in X .\ \sim P(x)$$

Similar arguments can be constructed to demonstrate the converse, and part (ii) of the theorem.   □

Notation

Further notational conveniences are as follows.

$$\forall m, n \in X . P(m, n)$$

means

$$\forall m \in X . \forall n \in X . P(m, n)$$

For example,

$$Prime(m) \triangleq m \neq 1 \wedge (\forall p, q \in \mathbb{N} . p * q = m \Rightarrow p = 1 \vee q = 1)$$

Similar notational shorthands are

$$\exists m, n \in X . P(m, n)$$
$$\forall m \in x, n \in y . P(m, n)$$
$$\exists x_1, x_2 \ldots x_n \in X . P(x_1, x_2 \ldots x_n)$$

and many other self-evident extensions.

Theorem 3.6

$$\exists x \in X . \exists y \in Y . P(x, y)$$
$$\equiv$$
$$\exists y \in Y . \exists x \in X . P(x, y)$$

and likewise

$$\forall x \in X . \forall y \in Y . P(x, y)$$
$$\equiv$$
$$\forall y \in Y . \forall x \in X . P(x, y)$$

Proof
Given

$$\exists x \in X . \exists y \in Y . P(x, y)$$

then there is some value val_1 of X and some value val_2 of Y such that

$$P(val_1, val_2)$$

Hence

$$\exists y \in Y . \exists x \in X . P(x, y)$$

An almost identical argument shows the converse. ☐
   Note however that it is not the case that

$$\forall x \in X . \exists y \in Y . P(x, y)$$
$$\equiv$$
$$\exists y \in Y . \forall x \in X . P(x, y)$$

For consider the proposition

$$\forall s \in Sons . \exists m \in Mothers . m \text{ is-mother-of } s$$

which 'translated into English' says that all sons have a mother, whereas

$$\exists m \in Mothers . \forall s \in Sons . m \text{ is-mother-of } s$$

would translate into: there is some mother who is the mother of all sons.
   There are certain other rules which enable one to factor out sub-expressions from quantified expressions as follows.

Theorem 3.7

1. $(\forall x \in X . P(x) \wedge Q(x)) \equiv (\forall x \in X . P(x)) \wedge (\forall x \in X . Q(x))$
2. $(\exists x \in X . P(x) \vee Q(x)) \equiv (\exists x \in X . P(x)) \vee (\exists x \in X . Q(x))$
3. $(\forall x \in \varnothing . P(x)) = \text{True}$
4. $(\exists x \in \varnothing . P(x)) = \text{False}$

where $\varnothing$ denotes the empty set.

Proof
1.   If

$$\forall x \in X . P(x) \wedge Q(x)$$

then, for arbitrarily chosen x

$$P(x) \wedge Q(x)$$

hence

$$P(x)$$

But x was arbitrarily chosen from X, so

$$\forall x \in X . P(x)$$

Likewise

$$\forall x \in X . Q(x)$$

hence

$$(\forall x \in X . P(x)) \wedge (\forall x \in X . Q(x))$$

For the converse the argument works in reverse. Given

$$(\forall x \in X . P(x)) \wedge (\forall x \in X . Q(x))$$

we can deduce that

$$\forall x \in X . P(x)$$

and hence that, for arbitrary x chosen from X

$$P(x)$$

Likewise for arbitrary x in X

$$Q(x)$$

and so, for arbitrary x in X

$$P(x) \wedge Q(x)$$

therefore

$$\forall x \in X . P(x) \wedge Q(x)$$

2. The same style of argument can be used to prove part 2.
    Alternatively de Morgan's laws (theorem 3.1) can be used to
    transform the statement into the form of part 1.
3, 4. The last two rules hold for all predicates $P(x)$: anything you can say
    about members of the empty set is true; on the other hand, you
    cannot find a member of the empty set (such that some property P is
    true of it).

Proofs Revisited

We have seen a number of theorems with different styles of proof in this
chapter. Firstly, we can substitute expressions in any formula with other
expressions which have the same value. If the formula has an arithmetic
sub-expression

$$x*(a+b)$$

we can substitute this using the distributive law and rewrite it as

$$x*a+x*b$$

We can carry out similar manipulation with logical expressions; so by
using de Morgan's theorem for instance, we can substitute

$$\sim(p \wedge q)$$

with

$$\sim p \vee \sim q$$

All such substitutions can be used as steps in the various kinds of
argument outlined in chapter 1 which constitute a proof.

To prove the identities of logical expressions we showed that their truth
tables were equivalent. This constitutes a proof because it amounts to
examining all possible values of the expressions; it is a proof by 'case
analysis'.

The third means of proof used in theorems 3.5, 3.6 and 3.7 is in the
style of 'Natural Deduction'. The main principle here is that one can
interchange between the structure of knowledge one has and the form of
the logical expression one may assert. Suppose one knows that the
proposition P is true, and also that the proposition Q is true. Then one
may deduce that

$$P \wedge Q$$

is true. Conversely if one knows that

$$P \wedge Q$$

as a single structured proposition is true, then one may deduce

P

and one may also deduce

Q

These rules are known as ' $\wedge$ -introduction' and ' $\wedge$ -elimination'
respectively. There are similar interchanges in respect of ' $\vee$ ' and ' $\sim$ ', and
one uses them almost unconsciously in the kinds of verbal argument used
in natural proofs. Another note is '*Modus Ponens*': if we know that P and
also that $P \Rightarrow Q$, then we can deduce Q. One may express such rules in the
following form

$$\frac{P, P \Rightarrow Q}{Q,}$$

and for $\wedge$ -introduction

$$\frac{P, Q}{P \wedge Q}$$

There are likewise exchanges related to the quantifiers. If we know that
$P(x)$ is true for any x chosen arbitrarily from set X, then we can deduce

$$\forall x \in X . P(x)$$

and conversely. Likewise if we know that there is some value *val* in the set
X such that $P(val)$, we can deduce

$$\exists x \in X . P(x)$$

and conversely. This kind of argument is used in the proofs of theorems
3.5, 3.6 and 3.7.

Exercise 3.6

As in exercise 3.5, express the following statements as predicates using
quantifiers

(a) London buses are red or green.
(b) No car more than twenty years old can go fast on cobbled roads.
(c) The houses in our street are red or green.
(d) Some London buses are red and some are green.

Example 3.10: Product Control System

It is time to consider some examples. In our product control system, each
item has an identification number and may be atomic or may in turn

consist of parts. In chapter 2 we said that for a practical example no part should be a member of itself, etc. Let the total set of products be *P*. Then

$$\sim\exists x{\in}P\,.\,x{\in}x$$

expresses this restriction. However, we want to go further, and insist that if we follow the tree of parts of any product, we eventually reach atomic components as indicated in figure 3.1.

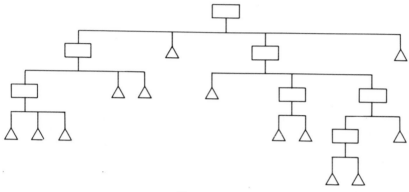

Figure 3.1

By 'walking through the tree' we eventually find all the leaves, for otherwise either the list of products is infinite or the structure is circular (see figure 3.2).

There are various ways of imposing this restriction. We may find the notion of a 'descendant' useful. We could define this recursively

$$x\,desc\,y \triangleq x{\in}y \vee (\exists z{\in}y\,.\,x\,desc\,z)$$

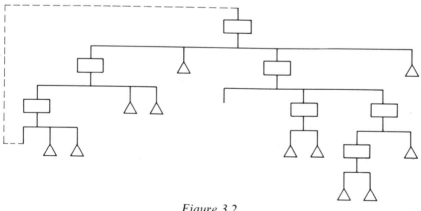

Figure 3.2

in which case '$x\ desc\ y$' means that x is a part of y or x is descended from a part of y. We can then stipulate

$$\forall x \in P. \sim x\ desc\ x$$

Alternatively, we may state this restriction in terms of sequences. Let $\{n\}$ denote the finite set of numbers i such that

$$0 \leqslant i < n$$

We can then say that there must be no sequence

$$p_0, p_1, \ldots, p_n$$

such that

$$\forall i \in \{n\} . p_i \in p_{(i+1)} \wedge (p_0 = p_n)$$

for any $n > 0$.

In later chapters we shall define the notions of sequences and indexed entities more rigorously. Also, in chapter 10 we go into more detail about defining functions, in particular recursive functions like *desc*.

Example 3.11: Chess Program

In our chess program we had the following sets

M: piece types
C: colours
F: files A . . . H
R: rows 1 . . . 8

For each $c \in C$, Pos_c is a collection of pieces from M each associated with a row and a file. Thus each member p_c of Pos_c has three attributes

$m(p_c) \in M$ the piece type
$f(p_c) \in F$ its file
$r(p_c) \in R$ its row

We must ensure that no two pieces of either colour are on the same square

$$\forall c1, c2 \in C . (\sim (\exists p_{c1} \in Pos_{c1}, p_{c2} \in Pos_{c2} . (f(p_{c1}) = f(p_{c2})$$
$$\wedge r(p_{c1}) = r(p_{c2}) \wedge (c1 \neq c2 \vee p_{c1} \neq p_{c2}))))$$

The last phrase $(c1 \neq c2 \vee p_{c1} \neq p_{c2})$ expresses the condition that the two actual pieces in question (p_{c1} and p_{c2}) are not the same.

For each piece type m there is a rule which dictates what are the possible moves for that piece. There are a number of possible ways of formulating this. For a rook we might have

$$m = \text{rook} \Rightarrow f' = f \vee r' = r$$

where f, r are the file and row of the original position of the piece and f', r' are the file and row of its new position, where $f, f' \in F$ and $r, r' \in R$; that is, where

$$m = m(p_c)$$
$$f = f(p_c)$$
$$r = r(p_c)$$

To take into account the other pieces, we might expand this expression

$$(m = \text{rook} \Rightarrow f' = f \vee r' = r) \wedge$$
$$(m = \text{bishop} \Rightarrow |f' - f| = |r' - r|) \wedge$$
$$(m = \text{queen} \Rightarrow f' = f \vee r' = r \vee |f' - f| = |r' - r|) \wedge$$
$$(m = \text{knight} \Rightarrow |f' - f| * |r' - r| = 2) \wedge$$
$$(m = \text{king} \Rightarrow |f' - f| = 1 \vee |r' - r| = 1)$$

where f, f' is a shorthand for the numerical value of f, f', such that $A = 1$, $B = 2$ etc. We could, with more precision, define a function $N(f)$, but since we have not dealt with functions yet, we shall use this less formal shorthand for the time being, just for this problem.

However, this is not complete because we have omitted pawns and we have not taken into account the obstruction of pieces. We may reformulate the above as follows

$$(((m = \text{rook} \wedge (f' = f \vee r' = r)) \vee$$
$$(m = \text{bishop} \wedge |f' - f| = |r' - r|) \vee$$
$$(m = \text{queen} \wedge (f' = f \vee r' = r \vee |f' - f| = |r' - r|))) \wedge$$
$$\text{unobstructed } (f', r', f, r))$$
$$\vee (m = \text{knight} \wedge |f' - f| * |r' - r| = 2)$$
$$\vee (m = \text{king} \wedge (|f' - f| = 1 \vee |r' - r| = 1))$$

where unobstructed $(f', r', f, r) \triangleq \forall p \in R, g \in F . \forall c \in C.$
$$\forall p_c \in Pos_c . (f(p_c) = g \wedge r(p_c) = p \wedge$$
$$(p = r = r' \wedge (g - f)(f' - g) \geqslant 0$$
$$\vee g = f = f' \wedge (p - r)(r' - p) \geqslant 0$$
$$\vee (r \neq r' \wedge f \neq f') \wedge (g - f)(r' - p)$$
$$= (p - r)(f' - g) \wedge (g - f)(f' - g) > 0)$$
$$\Rightarrow f(p_c) = f \wedge r(p_c) = r)$$

This expression states that the only piece in the original position Pos_c, $c = \text{Black, White}$, is the piece which is to be moved.

We have still not considered pawns or kings. Pawns move in different directions depending on their colour, can move two squares at first, can be promoted to other pieces, and other complications. Nor have we considered capture, not moving into check, and many other complications. However, this is probably sufficient analysis of this problem for the time being!

3.5 WORKSHOP EXERCISES

Exercise 3.7 (Workshop)

In the problem(s) which you stated at the end of chapter 1, and for which you identified and named sets at the end of chapter 2, consider if there are any conditions or constraints which apply to the members of those sets of data and use the notation of this chapter to express them.

Exercise 3.8 (Workshop)

In our main problem of the nuclear power station controller, we can envisage the following sets.

A set S of sensors. Each $s \in S$ is associated with a sensor type which is a member of

Types $= \{T, E, CT, CP, CF\}$ being Temperature, Energy output, Coolant temperature, Coolant pressure, Coolant flow rate. For each sensor type there is a range of values each of which forms a set as follows

Temp, a set of possible temperature values (for example, $-10°$ to $200°$) corresponding to sensor type T

En, a set of possible energy output values corresponding to sensor type E

C_{Temp}, a set of possible coolant temperature values, corresponding to sensor type CT

C_{Press}, a set of possible coolant pressure values, corresponding to sensor type CP

C_{Flow}, a set of possible coolant flow values, corresponding to sensor type CF.

The program which acts on all these also makes use of its own state recording the state of the system. This is likely to be a very large and complex set.

There is a set Ro of Rods

For each $r \in Ro$ there is a position pos_r which belongs to a set *Depths* of possible depths

$$Depths = \{d \in \mathbb{R} | \textit{min-depth} \leqslant d \leqslant \textit{max-depth}\}$$

Also associated with the rods there is a set of commands which the computer program can issue to each rod. The set of these commands may be

$$Comm = \{up, down\}$$

or

$$Comm = \{move \in \mathbb{R} | -\max \leqslant move \leqslant \max\}$$

The commands might conceivably include velocities and accelerations.

Associated with each sensor there is also a dial and a lamp: *Dials* is the set of dials and *Lamps* the set of lamps. Each member of *Dials* and *Lamps* is also associated with a member of *Types*. The program may send values to the dials, and the range of values is the same as that of the corresponding sensor. The program may also send a signal to any of the lamps. The values of the signal are members of the set

$$Signal = \{OK, Warn, Danger\}$$

Assuming that there is a dial and a lamp corresponding to each sensor, can you formulate a statement of the relationship between the value which the program reads from the sensor and that which it should send to the dial and lamp corresponding to it? The set V of VDUs do not correspond to particular sensors: a user may request on each that particular readings may be displayed. Try to formulate the set of requests (in abstract terms) and relate these to the messages which the program should display. The kinds of messages may belong to the set {Reading, Warning, Danger}. Assume that if any Danger Signals are present they must override the requests for reading, but that readings and 'OK' messages are displayed only in the absence of other requests.

Exercise 3.9 (Workshop)

There are many ways of setting up a diary system. One way is as follows: a diary may be considered to be a set of days, where each day is associated with a date and a set of events. Each event has a time, a duration and a description. The date is a member of the set of all possible dates, the time is a member of the set of possible times of day, and the duration is a positive number representing a period of some time (hours or minutes perhaps). All these sets, and typical members of them, may be given symbolic names.

Can you formulate a logical expression which is true of those events which occur (a) on a given day, and (b) at a given time on a given day? Suppose an event with its characteristics of time, duration and description is about to be entered on a specific date, can you formulate a logical expression which indicates whether or not the new event 'clashes' in time with any existing events in the diary? (This is more difficult.)

Exercise 3.10 (Workshop)

The space-borne computer system is likely to consist of a number of sub-systems

- Navigation
- Monitoring of observations
- Flight control
- Environmental control
- Communications

for example. Let us examine the Flight control sub-system. Certain information will be available, some from other sub-systems

- From the Navigation sub-system, the actual velocity, position, attitude and acceleration.
- From the Navigation sub-system and/or the Communication sub-system, the required velocity, position and attitude.
- From either the Navigation or the Observation system, the positions and velocities of close heavenly bodies, say the earth and the moon.
- From 'global data', the mass of the spacecraft and of the earth and moon.
- A set of motors, each with a range of 'burn rates', and durations of firing.

Each quantity, position, velocity, is a value which is a member of a possible range. Most of these are (for local space not affected by relativistic considerations) three-dimensional vectors

position	(x, y, z)	
velocity	$(\dot{x}, \dot{y}, \dot{z})$	
acceleration	$(\ddot{x}, \ddot{y}, \ddot{z})$	
attitude	(p, q, r)	$0 \leqslant p, q, r < 2\pi$ in radians
mass	m	$0 \leqslant m$ in kg
motors	a finite set	M
burn rate	$\dot{e}_f$	$0 \leqslant \dot{e}_f \leqslant max(f)$ where $f \in M$
duration	d_f	since different motors may have different maximum burn rates.

The acceleration of the spacecraft, $\ddot{s}$, is calculated from the set of burn rates, $\{\dot{e}_f | f \in M\}$, the mass m, the present position s, the position and mass of the earth $\overline{s_E}$, M_E, and of the moon $\overline{S_M}$, $\overline{m_M}$.

The necessary acceleration, $\ddot{\overline{SR}}$, to achieve the required position and velocity, $\overline{SR}$, $\dot{\overline{SR}}$, will be calculated from the present attitude, position, velocity and acceleration; this will be subject to constraints imposed by the maximum burn rates and the mass. From the required acceleration, $\ddot{\overline{SR}}$, and present acceleration, $\ddot{\overline{S}}$, can be calculated the burn rates $\{\dot{e}_f | f \in M\}$ and durations $\{d_f | f \in M\}$ for each motor $f \in M$. Without doing any celestial mechanics, that is by assuming that functions for calculating all quantities are given, can you express further constraints as logical expressions, for example to detect if after applying the calculated firings any further corrections are necessary, or to warn if the resulting positions, velocities etc. are different from the predicted quantities by excessive amounts?

Exercise 3.11 (Workshop)

Our fault-tolerant replicated real-time control system may consist of a finite set F of functional blocks, each of which is designed to perform the same function on some incoming data. Attached to each is a consistency checker, which carries out a supervisory check on the consistency of the output of the block with its input. We suppose that the checkers will not detect all faults however, and there is an arbiter which collects all the information output from the blocks and checkers (see figure 3.3). The input data d belongs to a range of data values D—that is, $d \in D$. Each $b \in F$

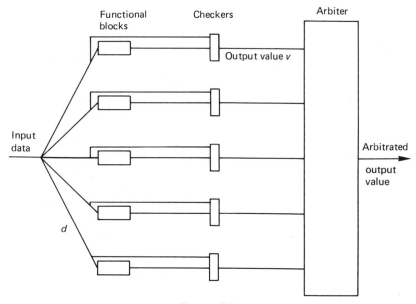

Figure 3.3

carries out a function f_b on d, producing an output $f_b(d)$ in the range E, so that $f_b(d) \in E$ for all $b \in F$. Each checker c_b in the set of checkers $\{c_b | b \in F\}$ carries out a checking function $ch(d, f_b(d))$ and produces an output value $v = f_b(d)$ if the check succeeds and an error value $v = \text{ERROR}$ if the check fails.

(a) Define the arbitrated output value by means of logical and set expressions.
(b) Define more formally the action of the checking function $ch(d, f_b(d))$.

SUMMARY OF RESULTS

Result	Reference
$(x \Rightarrow y) \equiv (\sim x \vee y)$	Example 3.1
$\sim(x \wedge y) \equiv (\sim x \vee \sim y)$	Theorem 3.1
$\sim(x \vee y) \equiv (\sim x \wedge \sim y)$	(De Morgan's laws)
$\wedge, \vee, \Leftrightarrow$ are commutative	Theorem 3.2
$\wedge, \vee, \Leftrightarrow$ are associative	Theorem 3.3
$\Rightarrow$ is not associative	Exercise 3.2
$\wedge$ is distributive with respect to $\vee$	Theorem 3.4
$\vee$ is distributive with respect to $\wedge$	
$\sim \sim x \equiv x$	Exercise 3.4
$(x \Leftrightarrow y) \equiv (x \Rightarrow y) \wedge (y \Rightarrow x)$	
$\forall x \in X . P(x) \equiv \sim \exists x \in X . \sim P(x)$	Theorem 3.5
$\sim \forall x \in X . \sim P(x) \equiv \exists x \in X . P(x)$	
$\exists x \in X . \exists y \in Y . P(x, y) \equiv \exists y \in Y . \exists x \in X . P(x, y)$	Theorem 3.6
$\forall x \in X . \forall y \in Y . P(x, y) \equiv \forall y \in Y . \forall x \in X . P(x, y)$	Theorem 3.6
$\forall x \in X . \exists y \in Y . P(x, y) \not\equiv \exists y \in Y . \forall x \in X . P(x, y)$	Theorem 3.6
$(\forall x \in X . P(x) \wedge Q(x)) \equiv (\forall x \in X . P(x)) \wedge (\forall x \in X . Q(x))$	Theorem 3.7
$(\exists x \in X . P(x) \vee Q(x)) \equiv (\exists x \in X . P(x)) \vee (\exists x \in X . Q(x))$	
$(\forall x \in \emptyset . P(x)) \equiv \text{True}$	
$(\exists x \in \emptyset . P(x)) \equiv \text{False}$	

Chapter 4 Sets and their Operations

NOTATION INTRODUCED

Notation	Concept	Reference
Card	cardinality (of finite sets)	Def. 4.1
0, 1, 2	the sets $\{\ \}, \{0\}, \{0, 1\}$	Section 4.1
ω	infinite set comprising natural numbers including 0	Def. 4.2
$\overline{\omega}$	Card ω	Def. 4.3
$\cong$	isomorphic, same cardinality	Def. 4.4
$\varnothing, \mathbf{0}, \{\ \}$	empty set	Section 4.1
$OP: A \times B \rightarrow C$	type, operation, arity, sort	Def. 4.6
$\cup$	union	Def. 4.7
$\cap$	intersection	Def. 4.8
	partition	Def. 4.9
$-, \setminus$	difference	Def. 4.10
$\subseteq$	subset	Def. 4.11
$=$	set equality	Def. 4.12
$\subset$	proper subset	Def. 4.13
$\times$	Cartesian product	Def. 4.15
$\underset{i}{\cup} A_i \quad \underset{i}{\cap} A_i$	distributed operations	Section 4.7
$\mathscr{P}$	powerset	Def. 4.16
$'$	complement	Def. 4.17
	Boolean Algebra	Def. 4.18

55

This chapter continues the study of sets with a number of basic operations one may carry out on them. Having now covered some elementary logic, we can define the meaning of these operations (which actually will probably be familiar to many readers).

4.1 CARDINALITY

Definition 4.1
We give an intuitive definition of cardinality for finite sets. The *cardinality* of a finite set S is the number of its members. The cardinality is denoted

Card (S)

Example 4.1
Most of the sets we have encountered so far are finite. The set S in examples 2.1 to 2.4 at the beginning of chapter 2 had five members, so its cardinality is 5, that is

Card $(S) = 5$

Likewise for the set

$T = \{34, \{34, \textit{farmer Giles's pig}, \text{"34"}, \textit{my house, the Bible}\}\}$

we have

Card $(T) = 2$

because T has two members.

Example 4.2: Personnel File

In our Personnel File example,

Card $(Alphabetic) = 52$
Card $(Ages)\quad = 50$
Card $(Grades)\quad = 12$
Card $(Salaries)\quad = 1\,000\,000$

The Empty Set

The set with no members, the empty set, is very important, just as the number zero is a 'very important number'. Because a set is identified by its members, there is just one unique empty set: like all other sets, it is the same no matter how it is defined

"The set of all Unicorns"

"The set of odd numbers divisible by 6"

"The set of all computers with a speed of more than 1000 MIPS"

(This last set may cease to be the empty set some time in the future!)

Notation

The empty set may be denoted in several ways

$$\mathbf{0} = \{\ \} = \varnothing$$

and

$$\text{Card } (\mathbf{0}) = 0$$

Also, for any x

$$x \in \{\ \}$$

is false.

Finite Sets

There is a standard collection of finite sets

$$\mathbf{0} = \{\ \} \qquad \text{Card } (\mathbf{0}) = 0$$
$$\mathbf{1} = \{0\} \qquad \text{Card } (\mathbf{1}) = 1$$
$$\mathbf{2} = \{0, 1\} \qquad \text{Card } (\mathbf{2}) = 2$$
$$\mathbf{3} = \{0, 1, 2\} \qquad \text{Card } (\mathbf{3}) = 3$$
$$\vdots$$
$$\mathbf{n} = \{0, 1, \ldots, n-1\} \qquad \text{Card } (\mathbf{n}) = n$$

If we have any set with set two members, we can 'pair off' its members with those of the set **2**; for example

$$\mathbf{2} = \{0, 1\}$$

or

$$\mathbb{B} = \{\text{True, False}\}$$
$$\mathbf{2} = \{0, 1\}$$

likewise

$$\{A, B\}$$
$$\{0, 1\}$$

etc., for any set with two members.

Much of the work with sets is concerned only with the structure of the sets, and not their contents. For sets the only intrinsic structure is the number of members—that is, the cardinality. In much formal work we may freely interchange such sets as $\{0, 1\}$, $\{True, False\}$, $\{Yes, No\}$. Such sets, which have the same cardinality, are 'isomorphic', which means 'of the same shape'. Isomorphism is however a more general concept: when applied to sets the only 'shape' of concern is the cardinality. Hence every finite set is isomorphic to the set **n**, with

$$\text{Card}\,(\overset{*}{S}) = \mathbf{n}$$

Infinite Sets

Consider the finite sets just defined

$$\mathbf{0, 1, 2, 3, 4, \ldots}$$

If we continue the progression to the limit, we obtain an infinite set.

Definition 4.2
$$\omega = \{0, 1, 2, 3, 4, 5, \ldots\}$$

all the natural numbers including 0. This set is often called $\aleph_0$.
Consider

$$\mathbb{N} = \{1, 2, 3, \ldots\}$$

and

$$evens = \{0, 2, 4, 6, 8, \ldots\}$$

We can set up one-to-one correspondences between the members of these sets

$$\omega = \{0, 1, 2, 3, 4, 5, \ldots, n, \ldots\}$$
$$\mathbb{N} = \{1, 2, 3, 4, 5, 6, \quad , n+1, \ldots\}$$

Similarly

$$\omega = \{0, 1, 2, 3, 4, \ldots, n, \ldots\}$$
$$evens = \{0, 2, 4, 6, 8, \ldots, 2n, \ldots\}$$

We wish to preserve the property that the cardinality of sets whose members can be paired one-to-one with each other is the same; hence the cardinality of all these is the same as that of ω.

Definition 4.3
Two sets *S1* and *S2* are said to be in *one–one correspondence*, or there is said to be a one–one correspondence between *S1* and *S2*, if there is a set of pairs *P* such that

$$P = \{(a, b) \,|\, a \in S1 \wedge b \in S2)$$

which has the uniqueness property:

$$\forall a \in S1 . \exists! b \in S2 . (a, b) \in P$$
$$\forall b \in S2 . \exists! a \in S1 . (a, b) \in P$$

Definition 4.4
Two sets $S1$, $S2$ are said to have the *same cardinality* if there is a one–one correspondence between them. One can then write

Card $(S1) =$ Card $(S2)$

Note that this definition applies to all sets, finite or infinite. In this case we also say that the sets are isomorphic and write

$$S1 \cong S2$$

A curious consequence of this is that an infinite set can be put in one–one correspondence with a 'proper subset' of itself—that is, a set whose members all belong to the infinite set but which does not contain all its members. Figure 4.1 illustrates this. By convention we define the cardinality of the infinite set ω as given below.

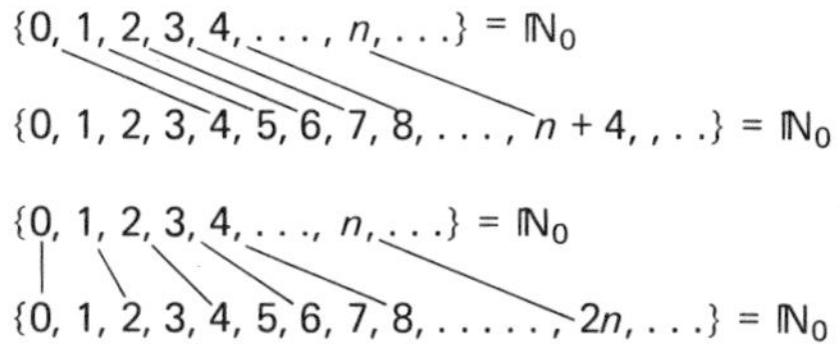

$$\{0, 1, 2, 3, 4, \ldots, n, \ldots\} = \mathbb{N}_0$$
$$\{0, 1, 2, 3, 4, 5, 6, 7, 8, \ldots, n + 4, , \ldots\} = \mathbb{N}_0$$

$$\{0, 1, 2, 3, 4, \ldots, n, \ldots\} = \mathbb{N}_0$$
$$\{0, 1, 2, 3, 4, 5, 6, 7, 8, \ldots, 2n, \ldots\} = \mathbb{N}_0$$

Figure 4.1

Definition 4.5

Card $(\omega) = \omega$

where ω is the smallest infinite number (which in set theory is usually called $\aleph_0$ pronounced 'Aleph-null', Aleph being the first letter of the Hebrew alphabet). Thus

Card $(\mathbb{N}_0) =$ Card $(\mathbb{N}) =$ Card(evens) $= \omega$

Application to Computing

In computing problems we do not actually have to manipulate infinite sets. However, we meet them when we try to generalise problems in order to simplify them: "The set of all possible character strings" and "The set of all possible ALGOL 60 programs" may be the kind of sets which we consider when generalising about the behaviour of programs.

If one wishes to discuss a large finite set it may be convenient to consider an infinite one. Analogous simplifications are done in numerical calculations. For example, Newton gave various finite iterative methods for finding approximations to square roots and integrals. The converse also applies. The following example illustrates this. Skip it if you have forgotten or have not covered the relevant calculus.

Example 4.3

If you have £100 invested for 10 years at 11 per cent compound interest, where the interest is accrued monthly, the sum at the end of the ten years will be

$$£100 + (1 + (11/1200)) ** 120$$

involving perhaps 119 multiplications (although some considerable arithmetic optimisation can be done to reduce this). An approximation is to pretend that the money is accruing continuously in an infinite number of infinitesimal steps. We then observe that the rate of increase of the sum of money x is

$$\frac{dx}{dt} = \frac{11}{100}x \qquad £/\text{year}$$

so that

$$\int_{x_0}^{x} \frac{100}{11x} dx = \int_{0}^{10} 1\, dt$$

Solving this gives $x_{10} = x_0 * e ** 1.1 = £100 * e ** 1.1$, which yields a quicker calculation (if one has a scientific calculator) than

$$£100 + (1 + 11/1200) ** 120$$

In the same kind of way we can often generalise about computing problems and the behaviour of programs, manipulating simply structured infinite sets instead of very large finite sets.

4.2 RUSSELL'S PARADOX

Implicit Definitions Again

Recall that a set can be defined implicitly by the expression

$$\{x \mid P(x)\}$$

meaning the set of all x such that $P(x)$ is true, where $P(x)$ is a Predicate function of x. We must be cautious using this notation, for we could define sets which do not exist. Recollecting the membership notation $\in$ and $\notin$, it is certainly true that most sets do not belong to themselves, for example

$$\{\ \}\notin\{\ \}$$

and

$$\{0\}\notin\{0\}$$

Let us consider all those sets which do not belong to themselves, that is

$$\{x\,|\,x\notin x\}$$

and call this set R. Now, does R belong to itself—that is, is the following true

$$R\in R\,?$$

If it is, then $R\in\{x\,|\,x\notin x\}$ so that by definition $R\notin R$. If it is not true, then R is not a set x such that $x\in x$—that is, $R\in R$ is not true, so that $R\notin R$ and we have a paradox, that is to say, we have a proposition which is neither uniquely true nor uniquely false.

This is called Russell's paradox, and there are various versions of it, some of them not immediately obvious. One homely version is the following

"In a village there is a barber. The barber shaves all those and only those, who do not shave themselves. Does the barber shave himself?"

Another version is

Consider all adjectives. We construct a skeleton sentence

'...' is a ... word

When an adjective is placed in both positions, a sentence is formed. The adjective is defined as 'Heterological' if the sentence so formed is false.

Thus

'Blue' is a blue word
'Hyphenated' is a hyphenated word

are both false, so that 'blue' and 'hyphenated' are heterological adjectives, whereas

'Polysyllabic' is a polysyllabic word

is true so that 'polysyllabic' is not a heterological word. *Question*: is 'heterological' a heterological word?

We avoid Russell's paradox by requiring that implicit set definitions take the form

$$\{x|x\in A \text{ and } P(x)\}$$

where A is a set already defined. This requirement may be relaxed where x is already in some known set such as the integers.

Ultimately, all infinite sets have to be defined by some means which does not involve explicit itemisation of their members.

Example 4.4

Some more examples of infinite sets are

$$\mathbb{N} = \text{Natural numbers}$$
$$= \{1, 2, 3, 4, \ldots\}$$
$$=_3 \{1, 10, 11, 100, 101, 110, 111, 1000, \ldots\}$$
$$= \{1, 2, 10, 11, 12, 20, 21, \ldots\}$$

(These are alternative representations of the same set, expressed in different bases.)

$$\mathbb{Q} = \text{Rational numbers}$$
$$= \{0, 1, -1, 2, -2, 3, -3 \ldots$$
$$1/2, -1/2, 3/2, -3/2, 5/2, \ldots,$$
$$1/3, -1/3, 2/3, -2/3, \ldots,$$
$$\ldots$$
$$\ldots\}$$

$$\mathbb{R} = \text{Real numbers}$$
$$= \{\text{co-ordinates of all the points on a line}\}$$

$$W = \{\text{all possible sequences which can be constructed from the alphabet } A, B, C, \ldots, Z\}$$

$$\text{ALGOL } 60 = \{\text{All legal ALGOL 60 programs}\}$$

An ALGOL 60 compiler has (among other things) to be able to distinguish members of ALGOL 60 from all other members of W' where

$$W' = \text{All sequences which can be constructed from the ALGOL 60 character set}$$

More generally a computer programming language is an infinite set of certain sequences from a particular character set which satisfy a certain prescription. This is called the 'syntax' of the language. A great deal of work and exploration has gone into various formalisms for defining the syntax of languages, called grammars. These grammars are stratified into well-defined levels of complexity. This is a subject of study in itself, about which many books have been written: see for example Backhouse (1979).

Example 4.5: Telecommunications

At a certain local telephone exchange, it is resolved that a computer-based subscriber accounts system should be set up. The sets involved are based on the following subsidiary sets

$$C = \{\text{“A”},\text{“a”},\text{“B”},\text{“b”},\ldots,\text{“Z”},\text{“z”},\text{“–”},\text{“.”},\text{“’”}\}$$
$$B = \{\text{“A”},\text{“a”},\ldots,\text{“z”},\text{“0”},\text{“1”},\ldots,\text{“9”},\text{“ ”},\text{“,”}\}$$
$$\mathbb{Z} = \text{the set of integers}$$

In the following X^* denotes the set of all sequences of members of the set X. We shall meet a more rigorous definition of this notation in chapter 5. The sets we require are as follows

S: a finite set of subscribers

C^*: the set of all possible names—that is, sequences of characters from the alphabet C

B^*: the set of all possible addresses—that is, sequences of characters from the alphabet B

TN: The set of all possible four-digit telephone numbers, being integers from 0000 to 9999—that is,
$$TN = \{n \mid n \in \mathbb{Z} \text{ and } n \geqslant 0 \text{ and } n < 10\,000\}$$

AC: The possible account position of subscribers; $AC = \mathbb{Z}$.

Our telephone exchange is small, since all telephone numbers in it consist of four digits. Subscribers can owe the telephone company a positive or negative amount.

Subscribers' names can consist of upper and lower case letters, hyphens, apostrophes and full stops. Their addresses may also contain digits, spaces and commas.

4.3 CONSTRUCTING NEW SETS FROM OLD

While the specific mechanisms discussed so far for constructing sets are useful, we soon meet their limitations. If we are not to be overwhelmed with lengthy definitions, we must find ways of using the definitions we have already framed, and constructing new sets in terms of those already formulated.

Operations and their Sorts

Just as arithmetic operations like $+$, $*$ and exp enable us to construct new numbers from a given set of numbers, so we have a number of operations on sets which produce new sets.

Definition 4.6: Types, Operations, Sorts, Arity

(i) A *type* is a set. Given a variable, its type is the set of values which it is permitted to adopt. Arguments of operations (see below) also have types which are the sets of values which the arguments may adopt.

  The types of variables in programming languages are typically (the set of) integers, (the set of) Booleans, (the set of) characters.

(ii) Operations have a fixed number of arguments and a result. An *operation* defines a correspondence between the values of its arguments and the value of its result. For any given combination of argument values there is a unique and unchanging value of its result.

(iii) The numbers of arguments possessed by an operation is its *arity*.

(iv) The types of the arguments and result of an operator are called *sorts*, and the collection of these sorts is called the *scheme* of the operator.

Notation

To express the scheme of an operator one uses special punctuation symbols: $\times$, $\rightarrow$ to separate the symbol standing for the operator, the types (scheme) of the respective arguments, and that of the result. See the next example.

Example 4.6

The sorts of the integer operations are written as follows

$$+ : \mathbb{Z} \times \mathbb{Z} \rightarrow \mathbb{Z}$$
$$- : \mathbb{Z} \times \mathbb{Z} \rightarrow \mathbb{Z}$$
$$* : \mathbb{Z} \times \mathbb{Z} \rightarrow \mathbb{Z}$$

The operator symbol is written before the ':'. Between the ':' and the '$\rightarrow$' we write the types of the operands, separated by '$\times$'. After the '$\rightarrow$' comes the type of the result. Thus the operations above take two integers as operands or 'arguments' and produce an integer result.

  In fact the symbols ':', '$\times$' and '$\rightarrow$' have rather more significance than mere punctuation, and we shall enlarge on this in chapters 5 and 6. For division we could write

$$/ : \mathbb{Z} \times \mathbb{Z} \rightarrow \mathbb{Q}$$

where $\mathbb{Q}$ is the set of rationals. (Actually this is not quite accurate because we have not defined the result if the second argument is zero. We should strictly show that the second argument belongs to the set consisting of all the integers excluding zero. We shall see shortly how to express this.) For operations on Reals we likewise have

$+:\mathbb{R}\times\mathbb{R}\rightarrow\mathbb{R}$

etc. Similarly, an if–then–else construction can be applied to any type T

if–then–else: Bool $\times\,T\times T\rightarrow T$

That illustrates the sorts of some of the operations with which we are familiar. Operations which produce new sets from old generally are of sort $S\times S\rightarrow S$, where S represents the set of all sets we are interested in. (This can usually be constructed by defining some collection as a universe and taking the set of all its subsets. The details need not concern us here.)
  We shall define

$\cup$	Union
$\cap$	Intersection
\ or $-$	Set difference
$\subseteq$	Subset
$\subset$	Proper subset
Card	Cardinality
$=$	Set equality
$\times$	Cartesian product

Definition 4.7: Union

$\cup:S\times S\rightarrow S$

where S denotes the sets of concern. Let P and Q be sets. Then

$P\cup Q=\{x\,|\,x\in P\vee x\in Q\}$

which means that $P\cup Q$ consists of elements which are either members of Q (or both) and no other elements.

Example 4.7

In our first example in chapter 2 where

$S=\{\text{farmer Giles's pig, 34, "34", my house, the Bible}\}$

and

$T=\{34,S\}$

then

$$S \cup T = \{\text{farmer Giles's pig, 34, "34", my house, the Bible, } S\}$$

Example 4.8: Time-table

Consider a school time-tabling problem. The general requirement is that a number of lessons should take place, where each lesson is a meeting of a number of pupils (possibly from different classes), a number of teachers (to allow for team teaching if necessary), a number of resources, a number of rooms (some may have removable partitions between them) etc.

Thus the principal data in the problem is five sets

> *Teachers*
> *Pupils*
> *Classes*
> *Resources*
> *Rooms*

Now each lesson can be represented by a set whose members belong to

Teachers ∪ *Pupils* ∪ *Resources* ∪ *Rooms*

(This is actually called a 'subset': more of this later.)

Example 4.9: Character Sets

In our character set example, we can more conveniently build up the classes of character sets as follows

> *Lower case* $= \{\text{"a", "b", "c", ..., "z"}\}$
> *Upper case* $= \{\text{"A", "B", "C", ..., "Z"}\}$
> *Alphabetic* $= Lower\ case \cup Upper\ case$
> *Digit* $= \{\text{"0", "1", ..., "9"}\}$
> *Alphanumeric* $= Alphabetic \cup Digit$
> etc.

Definition 4.8: Intersection

$$\cap : S \times S \rightarrow S$$

Let *P* and *Q* be sets. Then

$$P \cap Q = \{x | x \in P \land x \in Q\}$$

which means that $P \cap Q$ consists of those elements which are both members of *P* and members of *Q*.

Example 4.10

Back in example 4.7

$S = \{\text{farmer Giles's pig}, 34, \text{``34''}, \text{a house, the Bible}\}$

$T = \{34, S\}$

there is only one object which is a member both of S and T, namely 34.
Thus

$S \cap T = \{34\}$

Example 4.11: Product Control System

In our product example, two products were defined

$266\,982 = \{4011B, 3186, 7922C\}$

$37\,622A = \{3186, 2922C, 4099, 4011B\}$

then

$266\,982 \cap 37\,622A = \{4011B, 3186\}$

Example 4.12

In the time-tabling problem we divide the people in the school into a set
of non-intersecting groups called G, this set G being a set of sets such that
no pupil belongs to more than one set in G; that is

$\forall p \in R . \forall g1, g2 \in G . p \in g1 \wedge p \in g2 \Rightarrow g1 = g2$

or in other words

$\forall g1, g2 \in G . g1 \cap g2 = \emptyset \vee g1 = g2$

Either $g1$ and $g2$ have an empty intersection, or they are the same. This is
called a partition of a set, and pictorially can be depicted as in figure 4.2
or figure 4.3 (overleaf). All the points in the total area belong to just one
region. (We conveniently ignore points on the dividing lines!)

Definition 4.9

A Partition of a set S is a set T of sets such that

$(\forall x \in S . \exists s \in T . x \in S) \wedge (\forall s \in T . s \subseteq S \wedge s \neq \{\ \}) \wedge$

$(\forall x1, x2 \in T)(x1 \cap x2 = \{\ \} \vee x1 = x2)$

That is to say a partition of a set S is a set T of disjoint non-empty
subsets of S such that every member of S belongs to some (unique)
member of T.

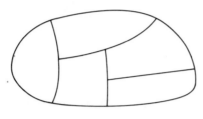 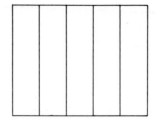

Figure 4.2   *Figure 4.3*

Example 4.13: Character Sets

Recall that we 'defined' a set

   ALGOL 60 = {all ALGOL 60 programs}

We could equally well have said

   FORTRAN = {all legal FORTRAN programs}

except that there are various dialects of FORTRAN!

   Let the dialects be $F1, F2, \ldots, Fn$. Then a subset common to all the dialects would be expressed

   $F1 \cap F2 \ldots \cap Fn$

Any member of this set would be a legal program according to all the dialects—a useful concept if one was trying to write portable software. Less useful of course if it turns out that

   $F1 \cap F2 \ldots \cap Fn = \varnothing$

In other words, there are no programs acceptable to all dialects! (Of course a further problem might be if the same program meant different things in different dialects—that is, if they had different semantics or behaved differently.)

Theorem 4.1

For finite sets there is a relationship between the cardinality of sets and that of their unions and intersections as follows

   $\text{Card}(S) + \text{Card}(T) = \text{Card}(S \cup T) + \text{Card}(S \cap T)$

Proof
Count the elements of $S \cup T$ by putting them into one–one correspondence with the elements of S and the elements of T. This will give a count $\text{Card}(S) + \text{Card}(T)$. However, every element in $S \cap T$ has been counted

twice (once as an element of S and once as an element of T) so

$$\text{Card}(S) + \text{Card}(T) = \text{Card}(S \cup T) + \text{Card}(S \cap T)$$

Definition 4.10: Difference

$$-: S \times S \rightarrow S$$

(The set difference symbol is often written '\'.) Let P and Q be sets. Then

$$P - Q = \{x \mid x \in P \wedge x \notin Q\}$$

Example 4.14
In example 4.8

$S = \{$farmer Giles's pig, 34, "34", my house, the Bible$\}$
$T = \{34, S\}$

then
$S - T = \{$farmer Giles's pig, "34", my house, the Bible$\}$
$T - S = \{S\}$

Example 4.15: Transactions on a File

This operation gives us flexibility in defining sets. For example, in our file-processing example we could have a restricted subset of the transactions defined for a certain class of users of the system

$\quad$ *Restricted Transactions = Transactions* $- \{\text{Purge}\}$

so that only users privileged to use the whole set of transactions would be able to purge the file.

Some Relationships

At this point it is worth pointing out some relationships with the logic ideas covered in chapter 3, and some resulting conclusions.

Observation 4.1

Let the set P be defined as

$$P = \{x \mid p(x)\}$$

where $p(x)$ is some predicate. Then the predicate

$$x \in P$$

is exactly equivalent to $p(x)$.

Proof
From the definition of

$$\{x|p(x)\}$$

it follows that

$$y\in\{x|p(x)\}$$

if and only if

$$p(y)$$

A similar result is given next.

Observation 4.2

$$P\cup Q = \{x|p(x) \vee x\in Q\}$$

Proof
This follows from the definition of $\cup$.

Observation 4.3

If likewise Q is defined

$$Q = \{x|q(x)\}$$

then

$$P\cup Q = \{x|p(x) \vee q(x)\}$$

and

$$P\cap Q = \{z|p(z) \wedge q(z)\}$$

(Recall that the name of the variable, x, z, is irrelevant since it is bound within the { } brackets.)

Proof
This follows from the definitions of $\cap$ and $\cup$.

Theorem 4.2

$$(P-Q)\cup(Q-P)=(P\cup Q)-(P\cap Q)$$

Proof
To show

$$(P-Q)\cup(Q-P)=(P\cup Q)-(P\cap Q)$$

we need to show

$$x \in (P-Q) \cup (Q-P) \Leftrightarrow x \in (P \cup Q) - (P \cap Q)$$

for arbitrary x. Assume

$$x \in (P-Q) \cup (Q-P)$$

This means that either

$$x \in (P-Q)$$

or

$$x \in (Q-P)$$

Consider the first case. Then

$$x \in P \text{ and not } x \in Q$$

Therefore

$$x \in P$$

and so

$$x \in P \cup Q$$

Also

$$x \notin Q$$

so

$$x \notin P \cap Q$$

hence

$$x \in (P \cup Q) - (P \cap Q)$$

The same argument applies in the second case where $x \in Q-P$.

The steps are easily reversed to give the implication in the opposite direction. □

Theorem 4.3

$$P-Q = P-(P \cap Q)$$

Proof
We have to show that

$$x \in P-Q \Leftrightarrow x \in P-(P \cap Q)$$

for arbitrary x. If

$$x \in P-Q$$

then

$x \in P$ and not $x \in Q$

From not $x \in Q$ it follows that

not $x \in (P \cap Q)$

hence

$x \in P$ and not $x \in (P \cap Q)$.

so that

$x \in P - (P \cap Q)$ □

4.4 VENN DIAGRAMS

The three operations we have met so far can usefully be illustrated with
Venn diagrams. In these diagrams sets are modelled by the collection of
points lying in a delimited area. In figure 4.4, all the points in the area
labelled S belong to the set S, all those in the area labelled T belong to

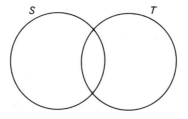

Figure 4.4

the set T. In each of the illustrations in figures 4.5–4.7 the shaded areas
represent the points in the set which is the result of the indicated
operation.

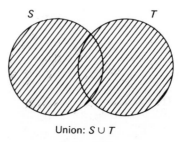

Union: $S \cup T$

Figure 4.5

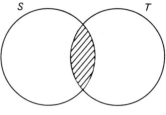

Intersection: $S \cap T$

Figure 4.6

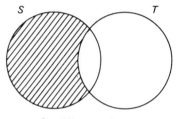

Set difference: $S - T$

Figure 4.7

4.5 MULTI-SORTED OPERATIONS

The next operations on sets which we shall discuss produce results which are not sets but, in this case, Booleans (True or False). These are examples of multi-sorted operators, whose arguments and results are of a variety of types.

Definition 4.11: Multi-sorted Operations
A Multi-sorted Operation is an operation whose arguments and result are not all of the same type.

Subsets

Definition 4.12: Subset

$$\subseteq : S \times S \to \mathbb{B}$$

Let P and Q be sets. Then

$$P \subseteq Q = (\forall x \in P . x \in Q)$$

Thus, P is a subset of Q means that all the members of P are members of Q.

Example 4.16

Consider S and T from example 4.14. Neither $S \subseteq T$ nor $T \subseteq S$.

Example 4.17: Car Supply Database

In the example relating to cars, it may be the case that each year different colours are brought out. If so, one may define

> *This year's colours* = {beige, maroon, black, puce}
>
> *Available colours this year*
> $$= \textit{This year's colours} \cup \textit{colours available last year}$$

in which case

> *This year's colours* $\subseteq$ *Available colours this year*

This suggests another relationship.

Theorem 4.4

$$P \subseteq Q \equiv P \cup Q = Q$$

Proof
We prove this by contradiction, taking two cases.
Case 1: suppose

$$P \cup Q \neq Q$$

Certainly all members of Q are also members of $P \cup Q$. So for $P \cup Q \neq Q$ there must be a member of $P \cup Q$ which is not a member of Q. Since $x \in P \cup Q$ if either $x \in P$ or $x \in Q$, there must be a member x of P which is not a member of Q. Hence

> not $P \subseteq Q$

Case 2: suppose

> not $P \subseteq Q$

then there is a member x of P such that $\sim x \in Q$. Since $x \in P$

$$x \in P \cup Q$$

Since we have a member x of $P \cup Q$ which is not a member of Q, it follows that

$$P \cup Q \neq Q$$

The combination of these two cases means that

$$P \subseteq Q \Leftrightarrow P \cup Q = Q \qquad \qquad \square$$

Theorem 4.5

$$(P \subseteq Q \wedge Q \subseteq R) \Rightarrow (P \subseteq R)$$

Proof
If

$$P \subseteq Q \wedge Q \subseteq R$$

then

$$x \in P \Rightarrow x \in Q \text{ and } x \in Q \Rightarrow x \in R$$

for any x. Hence

if $x \in P$ then $x \in Q$

and

if $x \in Q$ then $x \in R$

that is

$$P \subseteq R$$

so

$$P \subseteq Q \wedge Q \subseteq R \Rightarrow P \subseteq R \qquad \square$$

Definition 4.13: Equality

$$= : S \times S \to \mathbb{B}$$

Let P and Q be sets. Then

$$P = Q \Leftrightarrow (\forall x . x \in P \Leftrightarrow x \in Q)$$

That is

$$(\forall x . x \in P \Rightarrow x \in Q \wedge x \in Q \Rightarrow x \in P)$$

or

$$(\forall x . x \in P \Rightarrow x \in Q) \wedge (\forall x . x \in Q \Rightarrow x \in P)$$

hence

$$P \subseteq Q \wedge Q \subseteq P$$

In other words, all the members of P are members of Q and vice versa; or x is a member of P if and only if it is a member of Q.

Definition 4.14: Proper Subset

$$\subset : S \times S \to \mathbb{B}$$

Let P and Q be sets. Then

$$P \subset Q = (P \subseteq Q \wedge P \neq Q)$$

or equivalently

$$(\forall x . x \in P \Rightarrow x \in Q) \wedge (\exists x \in Q . x \notin P)$$

That is,

$$(\forall x . x \in P \Rightarrow x \in Q) \wedge \sim (\forall x . x \in Q \Rightarrow x \in P)$$

or

$$P \subseteq Q \wedge \sim Q \subseteq P$$

etc. In words, P is a subset of Q, but at least one member of Q is not a member of P. The proper subset operation in fact turns out not to be so useful as the ordinary subset operation.

Example 4.18: Car Supply Database

There are a number of instances of $\subset$ in our examples. With the car variants, if some new colours were actually introduced this year, then

This year's colours $\subset$ Available colours this year

Other examples of proper subsets can easily be found in the part numbers, the personnel file and the time-table examples.

Theorem 4.6

Like subsets, proper subsets display the property

$$(P \subset Q \wedge Q \subset R) \Rightarrow P \subset R$$

Proof
This is similar to that of theorem 4.5.

Example 4.19

An example from mathematical set theory is

$$\mathbb{N} \subset \mathbb{Z} \subset \mathbb{Q} \subset \mathbb{R}$$

Cardinality

The cardinality of a set S, denoted Card (S)—recall definitions 4.1, 4.4 and 4.5—can be considered as an operation which, given an argument which is a set, delivers its cardinality. Hence its sort is

$$\text{Card}: S' \rightarrow \mathbb{N}_0$$

where S' denotes finite sets. If P is a finite set, $\text{Card}(P)$ is the number of members of P.

Theorem 4.7

Certain relationships among the cardinalities of finite sets hold.

If $P \subseteq Q$, then $\text{Card}(P) \leqslant \text{Card}(Q)$
If $P = Q$, then $\text{Card}(P) = \text{Card}(Q)$
If $P \subset Q$, then $\text{Card}(P) < \text{Card}(Q)$

Proof
If $P \subseteq Q$ then each member of P is a member of Q. Count those members of Q which are also members of P. Suppose you have counted up to n. Then $n = \text{Card}(P)$. However, you have been counting members of Q; therefore

$$\text{Card}(P) < \text{Card}(Q)$$

Similar arguments apply to the second and third parts of the theorem. $\square$

4.6 CARTESIAN PRODUCTS

This is perhaps the most important and powerful of the set operations. The result of a Cartesian product operation is a set of ordered pairs. The pairs are written $\langle x, y \rangle$ where x belongs to the first argument set, and y belongs to the second.

Definition 4.15: Cartesian Product
$$\times : S \times S \rightarrow S$$

If P and Q are sets, then

$$P \times Q = \{\langle x, y \rangle \mid x \in P \wedge y \in Q\}$$

Note that $A \times (B \times C)$ and $(A \times B) \times C$ are not the same

$$A \times (B \times C) = \{\langle x, \langle y, z \rangle \rangle \mid x \in A \wedge y \in B \wedge z \in C\}$$

whereas

$$(A \times B) \times C = \{\langle \langle x, y \rangle, z \rangle \mid x \in A \wedge y \in B \wedge z \in C\}$$

However they are isomorphic, that is

$$A \times (B \times C) \cong (A \times B) \times C$$

Most of the time we can ignore this fact and write $A \times B \times C$. However it can be significant. For instance, suppose in an application we wanted to record a person's name assuming a first name, a second name, and a surname, then we might structure the record either as

> First_name
>
> Second_name
>
> Surname

corresponding to Name × Name × Name, or as

> Forenames = (First_name, Second_name)
>
> Surname

corresponding to (Name × Name) × Name.

Obviously the range of names we can record is identical in both cases, but we are conceptualising them in different ways.

This example also illustrates the difference between equality and isomorphism of sets. To identify a set is to identify its members—that is, to be able to represent them or give them names in some sense. Sets are equal if they contain the same named representatives; sets are isomorphic if the names in one can be unambiguously tied to the names in the other.

Example 4.20: Car Supply Database

A number of instances of Cartesian products (or cross products) can be found among our examples. In our car example, recall the definitions of the fundamental sets

> $B = \{\text{body styles}\}$
>
> $C = \{\text{colours}\}$
>
> $Eng = \{\text{different engine sizes}\}$
>
> $T = \{\text{interior trim styles}\}$

Then the total set of possible car variants is

> $B \times C \times Eng \times T$

A motivation for using the '×' symbol for Cartesian products is the following fact.

Theorem 4.8

For finite sets P and Q

$$\text{Card}\,(P \times Q) = \text{Card}\,(P) * \text{Card}\,(Q)$$

Proof

$P \times Q$ can be thought of as Card (Q) copies of P, namely

$$P \times \{\langle q1 \rangle\} \cup P \times \{\langle q2 \rangle\} \cup \ldots \cup P \times \{\langle qn \rangle\}$$

where $q1$ to qn are the different elements of Q. Since the sets $P \times \{\langle qi \rangle\}$ are pair-wise disjoint, the result follows. □

Example 4.21: Character Sets

In our character-set example, the set of all two-letter 'words' could be represented by

Alphabetic × *Alphabetic*

and three-letter words by

Alphabetic × *Alphabetic* × *Alphabetic*

etc.

Notation

$$S^n$$

denotes

$$\underbrace{S \times S \times S \ldots \times S}_{n \text{ times}}$$

Example 4.22: Part Numbers

In our example relating to part numbers, the part numbers themselves were of the form

Digit × *Digit* × *Digit* × *Digit*
Digit × *Digit* × *Digit* × *Digit* × *Digit*

etc. Representing

$$\underbrace{Digit \times \ldots \times Digit}_{n \text{ times}}$$

by *Digit*^{n}, a sequence of up to six digits could be defined as

$$DS = D^1 \cup D^2 \cup D^3 \cup D^4 \cup D^5 \cup D^6$$

and then finally allowing a letter suffix, we could define

Part No. $= DS \cup (DS \times Upper\ Case)$

which in words means that a part number consists of either a sequence of up to six digits, or a sequence of up to six digits followed by an upper-case letter.

Example 4.23: Personnel File

In our example from **Data Processing**, the personnel file, the information consisted of sets of names, ages, grades and salaries of employees. However, we did not show how a particular name could be associated with the age, grade etc. of a particular employee. With Cartesian products, we now have the equipment to do so. We can define the set of all possible employees

$$Employees = Names \times Ages \times Grades \times Salaries$$

Any employees can be characterised by a unique member of this set. A file of all the employees would be a subset of the set *Employees*.

Example 4.24: Time-table

In our time-tabling problem let

$$T = \{Teachers\}$$
$$C = \{Classes\}$$
$$L = \{Lesson\ types\}$$
$$P = \{Periods\}$$

A lesson could be represented by a member of the set

$$Lessons = T \times C \times L$$

so that each lesson is a unique combination of a teacher, a class, and a type of lesson. The events in the school time-table then consist of members of the set

$$P \times Lessons$$

or

$$P \times T \times C \times L$$

There are certain restrictions on the events in the time-table, such that no teacher can teach in more than one lesson in each period, and no class can have more than one lesson in one period. These constraints can be defined as follows. Let

$$Events \subseteq P \times T \times C \times L$$

then

(i) $\forall p \in P . \forall t \in T . \exists! c \in C . \exists! l \in L . \langle p, t, c, l \rangle \in Events$

(ii) $\forall p \in P . \forall c \in C . \exists! t \in T . \exists! l \in L . \langle p, t, c, l \rangle \in Events$

If in addition we consider classrooms, and suppose that each lesson takes place in a unique classroom, then

$$Lessons \subseteq T \times C \times L \times Rooms$$

and

$$Events \subseteq P \times T \times C \times L \times Rooms$$

Now only one lesson may take place in each room at once. We might stipulate

(iii) $\forall p \in P, r \in Rooms . \exists! t \in T . \exists! l \in L . \exists! c \in C . \langle p, t, c, l, r \rangle \in Events$

However, this would require that exactly one lesson takes place in each room during every period.

   If we wanted to allow some rooms to be occasionally empty, we would have to say

(iv) $\forall p \in P, r \in Rooms, t1, t2 \in T, l1, l2 \in L, c1, c2 \in C$

   $(\langle p, t1, c1, l1, r \rangle \in Events \wedge \langle p, t2, c2, l2, r \rangle \in Events)$
   $\Rightarrow (t1 = t2 \wedge c1 = c2 \wedge l1 = l2)$

which is saying that if two events take place at the same time (p) in the same room (r) then they are the same event by virtue of their other characteristics (teachers, class, subject) being the same.

   The same sort of formulation can be applied if one wishes to allow teachers and/or classes to have free periods—that is, to have periods during which no lesson takes place.

Example 4.25: Chess Program

In our chess-playing example, let us recall the sets we defined

$M = \{$Pawn, Rook, Knight, Bishop, King, Queen$\}$
$C = \{$Black, White$\}$
$F = \{$A,...,H$\}$ (files)
$R = \{1, 2 ... 8\}$ (rows)

Pos_c, the position associated with colour c, can now be more accurately described

$$Pos_c \subseteq F \times R \times M$$

Again, to express the constraints that at most one piece may occupy a square, we may stipulate

$$(\forall f \in F, r \in R, c1, c2 \in C, m1, m2 \in M$$
$$(\langle f, r, m1 \rangle \in Pos_{c1} \wedge \langle f, r, m2 \rangle \in Pos_{c2})$$
$$\Rightarrow (m1 = m2 \wedge c1 = c2)$$

Example 4.26: Telecommunications

Cartesian products can often be conveniently represented in high level programming languages by Data Structures (or Records in Pascal).

In our telephone exchange, we have

  S a set of subscribers
  C^* a set of sequences of characters being names
  B^* a set of sequences of other characters being addresses
  TN four-digit telephone numbers
  AC accounts

Now in fact TN can be defined further

$$TN = D \times D \times D \times D$$

where

$$D = \{\text{"0", "1", "2", "3", "4", "5", "6", "7", "8", "9"}\}$$

and the subscribers can be identified with their names, addresses and telephone numbers, so that we could define

$$S \triangleq C^* \times B^* \times TN$$

If we were writing a Pascal program to implement this telephone account system, we might have type declarations

    Subscriber = **record** (name string;
                           address string;
                           number telephone)

    Telephone = **array** $[1 \ldots \text{no digits}]$ digit;
    Digit = **set** $[0 \ldots 9]$

4.7 ASSOCIATIVITY

Recall definitions 3.6 and 3.7. We have the following.

Theorem 4.9

$\cup$ and $\cap$ are associative operations, that is

$$P \cup (Q \cup R) = (P \cup Q) \cup R$$
$$P \cap (Q \cap R) = (P \cap Q) \cap R$$

Thus we can drop the brackets without ambiguity, and write

$$P \cup Q \cup R$$
$$P \cap Q \cap R$$

respectively.

Proof

The proof follows quickly from the associativity of the logical operations $\vee$ and $\wedge$ (recall theorems 3.2 and 3.3)

$$P \cup (Q \cup R) = \{x \mid x \in P \vee x \in \{y \mid y \in Q \vee y \in R\}\}$$
$$= \{x \mid x \in P \vee (x \in Q \vee x \in R)\}$$
$$= \{x \mid (x \in P \vee x \in Q) \vee x \in R)\}$$

and since $\vee$ is associative

$$= \{x \mid x \in (P \cup Q) \vee x \in R\}$$
$$= (P \cup Q) \cup R$$

The same argument applies to $\cap$ (using the associativity of $\wedge$). $\quad\square$

The Cartesian product operation $\times$ can be regarded as associative if we consider $\langle \langle x, y \rangle, z \rangle$ to be the same as the ordered triple $\langle x, y, z \rangle$ and hence the same as $\langle x, \langle y, z \rangle \rangle$. However, when the distinction is important, $\times$ cannot be considered as associative.

Notation: Distributed Operators

For any associative operation, we can, purely for notational convenience, construct a 'distributed' notation; this is familiar for the arithmetic operations $+$ and $*$

$$\sum_{i=0}^{n} a_i = a_0 + a_1 + \ldots + a_n$$

$$\prod_{0 \leqslant i \leqslant n} a_i = a_0 * a_1 * \ldots * a_n$$

$$\sum_{i \in X} a_i = a_p + a_q + \ldots + a_r$$

where $X = \{p, q, \ldots, r\}$.

Likewise we can write notation such as

$$\bigcup_{i=0}^{n} A_i = A_0 \cup A_1 \ldots \cup A_n$$

$$\bigcup_{x \in S} A_x = A_{S_1} \cup A_{S_2} \ldots \cup A_{S_n} \qquad \text{where } S = \{s_1 \ldots s_n\}$$

$$\bigcap_{i=0}^{n} A_i = A_0 \cap A_1 \ldots \cap A_n$$

$$\bigcap_{x \in S} A_x = A_{S_1} \cap A_{S_2} \ldots \cap A_{S_n} \qquad \text{where } S = \{s_1 \ldots s_n\}$$

Similarly, where the context permits, we can write

$$\underset{i=0}{\overset{n}{\times}} A_i, \qquad \underset{x \in S}{\times} A_x$$

We can also write

$$\underset{i=1}{\overset{n}{\times}} A_i = A \times A \times \ldots \times A \qquad (n \text{ times})$$

where $n \geqslant 1$. Another accepted way of writing this is

$$A^n$$

Example 4.27: Telecommunications

In our telephone accounts system we can then define TN more succinctly as

$$TN = D^4$$

4.8 MORE RELATIONSHIPS

We saw how union ($\cup$) is intimately connected with Boolean disjunction or 'or' ($\vee$) and how intersection ($\cap$) is intimately connected with Boolean conjunction or 'and' ($\wedge$). There is a similar connection between subsets ($\subseteq$) and implication.

Earlier we saw how, strictly speaking all sets should be constructed from known sets or elements. So instead of constructions like

$$\{x \mid p(x)\}$$

we should write

$$\{x \mid x \in X \text{ and } p(x)\}$$

where X is some 'universe' for the present context.

Theorem 4.10

Let

$$P = \{x \mid x \in X \text{ and } p(x)\}$$

and

$$Q = \{x | x \in X \text{ and } q(x)\}$$

then

$$P \subseteq Q = (\forall x \in X)(p(x) \Rightarrow q(x))$$

Proof
$P \subseteq Q$ means that for arbitrary x

$$x \in P \Rightarrow x \in Q$$

but from the way P and Q are defined this can be rewritten as

$$p(x) \Rightarrow q(x)$$

However x was arbitrarily chosen from X so we can quantify thus

$$\forall x \in X . p(x) \Rightarrow q(x)$$

The converse argument applies.

4.9 POWERSETS

Definition 4.16: Powersets
The 'powerset' of a set S, written $\mathscr{P}(S)$, is defined as

$$\mathscr{P}(S) = \{x | x \subseteq S\}$$

that is, the set of all subsets of S. If S is a set, $\mathscr{P}(S)$ as defined above will also be a set—that is, we are not in danger of encountering paradoxes such as Russell's paradox (see section 4.2).

Example 4.28

Recall the set

$$\mathbf{1} = \{0\}$$

What are all the subsets of **1**? The empty set $\{\ \}$ is a subset of any set: all its members (there are none) are members of any given set. So

$$\{\ \} \subseteq \{0\}$$

Also **1** is a subset of **1**: all the members of **1** are members of **1**. So

$$\{0\} \subseteq \{0\}$$

Intuitively we may conclude that there are no other subsets of **1**. The set of all subsets of **1** is the powerset of **1**, that is

$$\mathscr{P}(\mathbf{1}) = \{\{\ \}, \{0\}\}$$

What are the subsets of **2**? Recall that

$$\mathbf{2} = \{0, 1\}$$

Then

$$\mathscr{P}(\mathbf{2}) = \{\{\ \}, \{0\}, \{1\}, \{0, 1\}\}$$

Likewise

$$\mathscr{P}(\mathbf{3}) = \{\{\ \}, \{0\}, \{1\}, \{2\}, \{1, 2\}, \{0, 2\}, \{0, 1\}, \{0, 1, 2\}\}$$

and so on.

Theorem 4.11

For a finite set S of cardinality n, the cardinality of $\mathscr{P}(S)$ is 2^n.

Proof
We can give a unique representation of any subset of S as follows.
Consider a string of n binary digits where each digit is associated with a
particular member of S. A subset can be represented by a unique value of
the n-digit string, with the ith place set to 1 if the corresponding member
of S belongs to the subset and 0 otherwise. Any such string represents a
subset of S and the representations are unique for a given correspondence
of positions in the string to members of S. Hence the set of all such strings
is in one–one correspondence to the set of all subsets of S, and so has the
same cardinality. However the number of values of an n-digit binary string
is 2^n. □

Exercise 4.1

What is the cardinality of the empty set $\mathbf{0}$?
What is the cardinality of its powerset?
What is its powerset?
What is $\mathscr{P}(\mathscr{P}(\mathbf{0}))$?
And $\mathscr{P}(\mathscr{P}(\mathscr{P}(\mathbf{0})))$?
What is Card $(\mathscr{P}(\mathscr{P}(\ldots \mathscr{P}(\mathbf{0}))\ldots))$?

$$\underbrace{\hphantom{\mathscr{P}(\mathscr{P}(\ldots \mathscr{P}(\mathbf{0}))\ldots)}}$$

n '$\mathscr{P}$'s

Let us now consider all the subsets of S where $S = \{a, b, c\}$—that is,
$\mathscr{P}(S)$—and construct a table for union, intersection, complement, and
subset.

	{a,b,c},	{a,b},	{a,c},	{a},	{b,c},	{b},	{c},	{ }
∪	S1	S2	S3	S4	S5	S6	S7	S8
S1	S1	S1	S1	S1	S1	S1	S1	S1
S2	S1	S2	S1	S2	S1	S2	S1	S2
S3	S1	S1	S3	S3	S1	S1	S3	S3
S4	S1	S2	S3	S4	S1	S2	S3	S4
S5	S1	S1	S1	S1	S5	S5	S5	S5
S6	S1	S2	S1	S2	S5	S6	S5	S6
S7	S1	S1	S3	S3	S5	S5	S7	S7
S8	S1	S2	S3	S4	S5	S6	S7	S8
∩	S1	S2	S3	S4	S5	S6	S7	S8
S1	S1	S2	S3	S4	S5	S6	S7	S8
S2	S2	S2	S4	S4	S6	S6	S8	S8
S3	S3	S4	S3	S4	S7	S8	S7	S8
S4	S4	S4	S4	S4	S8	S8	S8	S8
S5	S5	S6	S7	S8	S5	S6	S7	S8
S6	S6	S6	S8	S8	S6	S6	S8	S8
S7	S7	S8	S7	S8	S7	S8	S7	S8
S8	S8	S8	S8	S8	S8	S8	S8	S8

Definition 4.17

The complement of a set R is the set of members of the 'universe' (in the above example S) which are not members of R—that is

$$R' = \{x \mid x \in S \land x \notin R\}$$

It follows that for $\mathscr{P}(S)$, the table of ' is

S1	S2	S3	S4	S5	S6	S7	S8
S8	S7	S6	S5	S4	S3	S2	S1

Finally

⊆	S1	S2	S3	S4	S5	S6	S7	S8
S1	T	T	T	T	T	T	T	T
S2	F	T	F	T	F	T	F	T
S3	F	F	T	T	F	F	T	T
S4	F	F	F	T	F	F	F	T
S5	F	F	F	F	T	T	T	T
S6	F	F	F	F	F	T	F	T
S7	F	F	F	F	F	F	T	T
S8	F	F	F	F	F	F	F	T

where T and F stand for true and false respectively, and each cell represents the truth value of the statement that the subset of S in whose

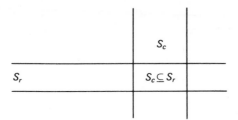

Figure 4.8

column it lies is a subset of the subset of S in whose row it lies. See figure 4.8.

Now finally we can construct the tables of operations on the predicate functions on S: the operations being $\lor$, $\land$, $\sim$ and $\Rightarrow$ extended to predicate functions in the following way

$$\forall x \in S.\,((f \lor g)(x) \triangleq f(x) \lor g(x))$$

Thus the operation $\lor$ when applied to two predicate functions gives a new predicate function whose value when applied to each x in S is as defined above. The other extended operations are as follows

$$\forall x \in S.\,((f \land g)(x) \triangleq f(x) \land g(x))$$
$$\forall x \in S.\,((\sim f)(x) \triangleq \sim(f(x)))$$
$$\forall x \in S.\,((f \Rightarrow g)(x) \triangleq (f(x) \Rightarrow g(x)))$$

Copious brackets have been included to render absolutely unambiguous the parsing of these definitions so that it is clear, for example, that $(f \Rightarrow g)$ is a new function!

Now the table for $\lor$ when applied to the predicate functions on S, as defined earlier, is as follows

$\lor$	$P1$	$P2$	$P3$	$P4$	$P5$	$P6$	$P7$	$P8$
$P1$	$P1$	$P1$	$P1$	$P1$	$P1$	$P1$	$P1$	$P1$
$P2$	$P1$	$P2$	$P1$	$P2$	$P1$	$P2$	$P1$	$P2$
$P3$	$P1$	$P1$	$P3$	$P3$	$P1$	$P1$	$P3$	$P3$
$P4$	$P1$	$P2$	$P3$	$P4$	$P1$	$P2$	$P3$	$P4$
$P5$	$P1$	$P1$	$P1$	$P1$	$P5$	$P5$	$P5$	$P5$
$P6$	$P1$	$P2$	$P1$	$P2$	$P5$	$P6$	$P5$	$P6$
$P7$	$P1$	$P1$	$P3$	$P3$	$P5$	$P5$	$P7$	$P7$
$P8$	$P1$	$P2$	$P3$	$P4$	$P5$	$P6$	$P7$	$P8$

Observe the similarity with the table for $\cup$ on the powerset members of S. This illustrates further the intimate relationship between subsets and predicates, and between $\cup$ and $\lor$. Constructing the tables for $\land$, $\sim$ and

⇒ for the predicates of the set is left as an exercise for the reader: observe again the relationship between ∩ and ∧,′ and ∼, ⊆ and ⇒.

Boolean Algebra

This is something of a postscript to the preceding section, and can be omitted if desired since it is not an essential part of the treatment.

Definition 4.18: Boolean Algebra
A Boolean Algebra is a set C with a binary operation ∩ and a unary operation ′ which satisfy the following axioms

(1) $x \cap y = y \cap x$ for all $x, y \in C$
(2) $(x \cap y) \cap z = x \cap (y \cap z)$ for all $x, y, z \in C$
(3) If $x \cap y' = z \cap z'$ then $x \cap y = x$ for all $x, y, z \in C$
(4) If $x \cap y = x$ then $x \cap y' = z \cap z'$ for all $x, y, z \in C$

Exercise 4.2

There are various models which satisfy these axioms and which therefore are Boolean Algebras. The most well known one is {true, false} with the operations ∧ and ∼. Check using the truth tables that the four axioms hold. Another model is the powerset of any set S, with the operations of ∩ and ′. Again, using the three-member set S described in the foregoing sections, one can verify that the intersection and complement operations satisfy the axioms for a Boolean Algebra.

4.10 EXAMPLES

Example 4.29: Bus Routes

In our bus route problem, a route could be regarded as a subset of the set of districts D, so that the set of all routes *Busr* would satisfy

$$Busr \subseteq \mathscr{P}(D)$$

or equivalently

$$Busr \in \mathscr{P}\mathscr{P}(D)$$

However, this would ignore the ordering of the districts in a route, and would ignore the possibility of a bus route visiting the same district twice.

Example 4.30: Chess Program

In our chess-playing example, we observed that each position Pos_c was a subset of $F \times R \times M$ where F is the set of files, R the set of rows, and M the set of pieces. Thus

$$Pos_c \in \mathscr{P}(F \times R \times M)$$

Example 4.31: Telecommunications

In our telephone exchange example, we could elaborate the problem a little by allowing a set of facilities which each subscriber could have

$F = \{$Push-button dialling,
        Extra extensions,
        Abbreviated dialling,
        Facsimile equipment,
        ...
        ...$\}$

Each subscriber may have none or several or even all of these facilities. So, in addition to the name, address and telephone number which identified a subscriber, we could add a subset of the facilities. Thus

$$S \subseteq C^* \times B^* \times TN \times \mathscr{P}(F)$$

or equivalently

$$S \in \mathscr{P}(C^* \times B^* \times TN \times \mathscr{P}(F))$$

Exercise 4.3 (Workshop)

Again take your project problem(s) which you have gradually formalised from chapter 1. Can any of the data sets be more exactly defined using unions, intersections, Cartesian products? Can relationships between the data sets be defined using subsets or powersets?

Exercise 4.4 (Workshop)

In the nuclear power station exercise, consider how the problem definition can be likewise refined using the concepts from this chapter. In particular consider the data output from a sensor being the union of the range of the sensor and an error signal resulting from a failure of the sensing mechanism. Can other aspects of the problem be enlarged to cater for failure of parts of the system?

Exercise 4.5 (Workshop)

Refine the sets of data in the personal-computer diary/reminder problem. For example, the date may consist of a Cartesian product $D \times M \times Y$ where $D = \{\text{days of the week}\}$, $M = \{\text{months of the year}\}$, $Y = \{\text{years}\}$. Similarly the time of day may be structured, and also much of information forming the format of requests and issued information may consist of members of Cartesian products, unions and so on.

SUMMARY OF RESULTS

Result	Reference
$\text{Card}\,(S \cup T) + \text{Card}\,(S \cap T) = \text{Card}\,S + \text{Card}\,T$	Theorem 4.1
$y \in \{x \mid p(x)\} \equiv p(y)$	Observation 4.1
$\{x \mid p(x)\} \cup Q = \{x \mid p(x) \vee x \in Q\}$	Observation 4.2
$\{x \mid p(x)\} \cup \{x \mid q(x)\} = \{x \mid p(x) \vee q(x)\}$	Observation 4.3
$\{x \mid p(x)\} \cap \{x \mid q(x)\} = \{x \mid p(x) \wedge q(x)\}$	
$(P - Q) \cup (Q - P) = (P \cup Q) - (P \cap Q)$	Theorem 4.2
$P - Q = P - (P \cap Q)$	Theorem 4.3
$P \subseteq Q \equiv P \cup Q = Q$	Theorem 4.4
$(P \subseteq Q \wedge Q \subseteq R) \Rightarrow (P \subseteq R)$	Theorem 4.5
$(P \subset Q \wedge Q \subset R) \Rightarrow (P \subset R)$	Theorem 4.6
If $P \subseteq Q$ then $\text{Card}\,P \leqslant \text{Card}\,Q$ etc.	Theorem 4.7
$\text{Card}\,(P \times Q) = \text{Card}\,P * \text{Card}\,Q$	Theorem 4.8
$\cup, \cap$ are associative	Theorem 4.9
$\{x \mid p(x)\} \subseteq \{x \mid q(x)\} \equiv p(x) \Rightarrow q(x)$	Theorem 4.10
$\text{Card}\,\mathscr{P}(S) = 2^{\text{Card}\,S}$	Theorem 4.11

PART II: Data Type Construction

In chapter 4 we saw how to construct new sets out of given sets, using the operations 'intersection', 'union' and 'Cartesian product'. This gave us some expressive power in that we can now define sets by building on those we have already defined. However, the ways in which we can build up further sets are still rather limited. We need some more notational apparatus to give us the expressive flexibility in order to capture the essential characteristics of the data encountered in information-processing applications.

In Part II we introduce three more concepts which enable us to relate sets to each other in further ways. These are

> Tuples
> Mappings and functions
> Relations

We shall then be able to define abstract models of practically all information structures we may encounter in software engineering applications. Having completed Part II the reader will be well equipped to construct and capture abstract formulations of many Data Types, both simple and complex.

Chapter 5 Tuples, Products and Sums

NOTATION INTRODUCED

Notation	Concept	Reference
$\langle \ldots \rangle$	tuple	Section 5.1
'n-tuple'	tuple of length n	Section 5.1
$\prod_{i \in F} A_i$	distributed Cartesian product, alternative to $\underset{i \in F}{\times} A_i$	Section 5.1
A^*	$\underset{i \geqslant 0}{\cup} A^i$ (Kleene star)	Def. 5.2
A^+	$\underset{i > 0}{\cup} A^i$	Def. 5.3
$\oplus$	disjoint union	Def. 5.4
$\underset{i \in I}{\oplus} A_i$	distributed sum	Section 5.2
heterogeneous products and tuples	$\underset{i \in F}{\times} A_i$ where the A_i values are not the same; their members	Def. 5.5
'represents'		Def. 5.6
'string, sequence, list'		Def. 5.7
$\parallel$	concatenation	Def. 5.8
hd	head of a tuple	Def. 5.9
tl	tail of a tuple	Def. 5.10
len	length of a tuple	Def. 5.11
$s(i)$	indexing a tuple	Def. 5.12

5.1 TUPLES

A tuple is the general name for a pair, a triple, a quartuple, quintuple, sextuple, etc. It is a generalisation of a Cartesian product, which we met in chapter 4. In section 4.5 we met the notation for a distributed Cartesian product

$$\underset{i \in I}{\times} A_i$$

where I is some finite index set.

Definition 5.1
A *tuple* is any member of some distributed Cartesian product

$$\underset{i \in I}{\times} A_i$$

where the A_i are sets defined for $i \in I$, where I is some finite index set.

Notation

In particular if the cardinality of I is n, the members of the above product are called *n-tuples*. When $n = 2$, we talk of *pairs*; when $n = 3$, of *triples*; etc.

Example 5.1

The pairs $\langle a, b \rangle$ of elements $a \in A, b \in B$, form the Cartesian product $A \times B$, and can be considered as co-ordinates for the product. See figure 5.1.

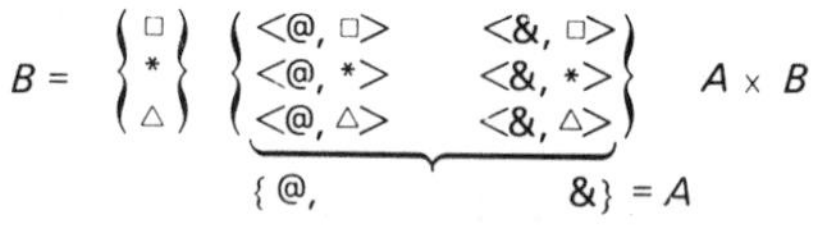

$$B = \begin{Bmatrix} \square \\ * \\ \triangle \end{Bmatrix} \quad \begin{Bmatrix} \langle @, \square \rangle & \langle \&, \square \rangle \\ \langle @, * \rangle & \langle \&, * \rangle \\ \langle @, \triangle \rangle & \langle \&, \triangle \rangle \end{Bmatrix} \quad A \times B$$
$$\underbrace{\{ @, \qquad \& \}}_{} = A$$

Figure 5.1

Example 5.2: Car Supply Database

Recalling our car part numbers example, figure 5.2 illustrates the use of Cartesian products. Figure 5.3 illustrates a 'degenerate' example.

Example 5.3: The Euclidean Plane

The real numbers $\mathbb{R}$ give a name for every point on the line. In analytical geometry we learn that $\mathbb{R} \times \mathbb{R} = \mathbb{R}^2$ provides co-ordinates for every point in

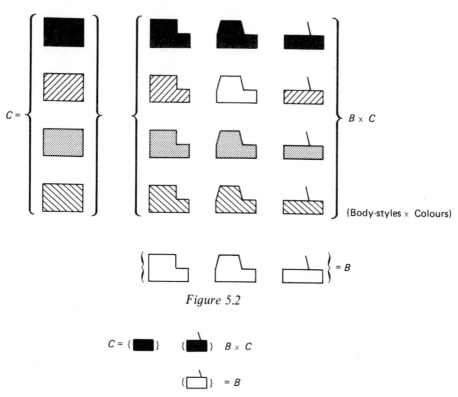

Figure 5.2

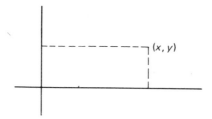

"You can have any colour as long as it is Black"

Figure 5.3

Figure 5.4

the plane—see figure 5.4—and that $\mathbb{R}^n$ gives a co-ordinate system for n-dimensional space.

Example 5.4: State Machines

Electronic engineering systems, especially those encountered in telecommunications, are often described by enumerating a set of states, and defining conditions under which the system will transit from one state to another. If two systems with 'state spaces' S_1 and S_2 respectively interact with each other, the composite system will have a state space $S_1 \times S_2$.

If the system above has many, say n, components with state spaces S_i, $1 \leqslant i \leqslant n$, the composite system has a state space

$$\underset{1 \leqslant i \leqslant n}{\times} S_i$$

A member of this composite space will be an n-tuple of form

$$\langle s_1, s_2, \ldots, s_i, \ldots, s_n \rangle$$

where each $s_i \in S_i$.

Example 5.5: Words over an Alphabet

An n-letter word can be considered as an n-tuple.

If we have an alphabet A then A^n is the set of all n-letter words. Consider A^2 illustrated in figure 5.5. The set of all one-letter words is A^1,

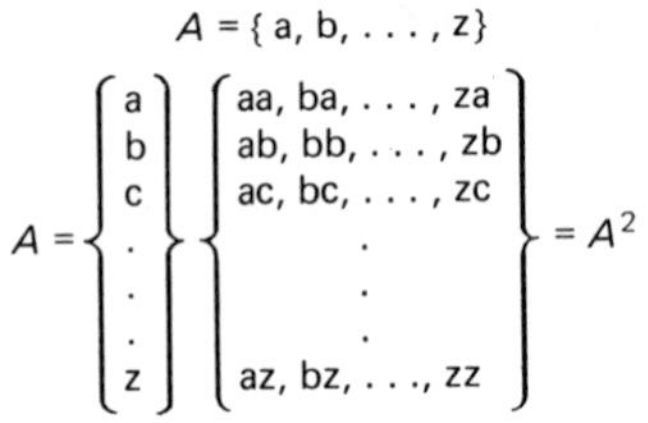

$$A = \{a, b, \ldots, z\}$$

$$A = \left\{ \begin{array}{c} a \\ b \\ c \\ . \\ . \\ . \\ z \end{array} \right\} \left\{ \begin{array}{c} aa, ba, \ldots, za \\ ab, bb, \ldots, zb \\ ac, bc, \ldots, zc \\ . \\ . \\ az, bz, \ldots, zz \end{array} \right\} = A^2$$

Figure 5.5

two letter words is A^2, three letter words is A^3 etc. The set of all words of 1, 2 or 3 letters is

$$A^1 \cup A^2 \cup A^3$$

The set of all words of up to and including n letters, including the empty word of zero letters, is

$$\underset{0 \leqslant i \leqslant n}{\cup} A^i$$

Note that the empty tuple $\langle \ \rangle$ is the unique member of A^0, which thus has one member and has cardinality 1. In fact the relationship between cardinality and Cartesian products is reminiscent of the symbol $\times$ used for Cartesian product.

Theorem 5.1

For finite sets A, B, A_i, F

$$\text{Card}\,(A \times B) = \text{Card}\,A * \text{Card}\,B$$

$$\text{Card}\, \underset{i \in F}{\times}\, A_i = \prod_{i \in F} A_i$$

where $\prod$ represents distributed arithmetic product.

Proof
Given $a \in A$ then there are (Card B) distinct tuples

$$\langle a, x \rangle$$

where $x \in B$. Hence, since there are (Card A) distinct members of A, there are Card $A \times$ Card B distinct tuples $\langle y, x \rangle$ where $y \in A, x \in B$.

The proof of the second statement follows by induction over the cardinality of F.

Notation

The distributed Cartesian product is sometimes written

$$\prod_{i \in F} A_i$$

Definition 5.2
Lastly, we generalise the n-tuple as follows. The set of all words of any length is

$$\underset{0 \leqslant i}{\cup}\, A^i$$

and is written

$$A^*$$

Thus

$$A^* = \{\langle\ \rangle, a, b, \ldots, z, aa, ab, \ldots, zz, aaa, aab, aba, \ldots, aba, \ldots, zzz,$$
$$aaaa, \ldots, aaaaa, \ldots, \ldots \text{indefinitely}\}$$

This notation is sometimes known as the 'Kleene star'.

Definition 5.3
A^+ is the set of words which does not contain the empty word $\langle\ \rangle$

$$A^+ = \underset{1 \leqslant i}{\cup}\, A^i$$

Note that $A^1 = \{\langle a \rangle | a \in A\} \neq A$ but clearly A^1 and A are isomorphic, and thus the distinction is often ignored.

Exercise 5.1

'Real' numbers in a programming language are represented by sequences of characters as follows

> a non-null sequence of digits followed by
> a point '.' followed by
> a possibly null sequence of digits followed by
> 'E' followed by '+' or '−' followed by
> a non-null sequence of digits

Express the set of possible formulations of these numbers as a Cartesian product.

5.2 SUMS AND PRODUCTS

This section can be deferred until after chapter 6 when homomorphisms and injections will have been considered more thoroughly, if desired.

Cartesian products are often simply called 'products'. Another related construction is the (Cartesian) 'sum' (or 'disjoint union' or 'coproduct'). This is written $\oplus$ (or sometimes simply $+$).

Definition 5.4: Disjoint Union
Given sets $S1$, $S2$, then

$$S1 \oplus S2$$

is the union of disjoint *copies* of $S1$ and $S2$. If the sets $S1$ and $S2$ are themselves disjoint, then $S1 \oplus S2 \cong S1 \cup S2$.

Associated with disjoint union is a pair of 'injection mappings', i_L and i_R

$$i_L : S1 \rightarrow S1 \oplus S2$$

$$i_R : S2 \rightarrow S1 \oplus S2$$

and are such that

$$S1 \oplus S2 = rng\ i_L \cup rng\ i_R$$

and

$$rng\ i_L \cap rng\ i_R = \{\ \}$$

The fact that i_L and i_R are injections means that they are one-to-one. It follows that

$$x \in S1 \oplus S2 \Leftrightarrow \exists y \in S1 . i_L(y) = x \vee \exists y \in S2 . i_R(y) = x$$

The symbol $\oplus$ denotes an operation on two sets which produces a third set containing distinct copies of members of both the operand sets. It is called variously 'Disjoint Union', 'Sum' and 'Coproduct'. It captures a naive idea of 'adding two sets together'. The following identities may be noted.

Theorem 5.2

If $S1$, $S2$ and $S3$ are pair-wise disjoint, then

(1) Card $(S1 \oplus S2) =$ Card $(S1) +$ Card $(S2)$

(2) $\varnothing \oplus S \cong S \oplus \varnothing \cong S$

(3) $S1 \oplus S2 \cong S2 \oplus S1$

(4) $(S1 \oplus S2) \oplus S3 \cong S1 \oplus (S2 \oplus S3)$

Proof

(4) follows from the associativity of $\vee$.

(3) follows from the commutativity of $\vee$.

(2) is proved as follows

$$x \in \varnothing \oplus S \Leftrightarrow \exists y \in \varnothing . i_L(y) = x \vee \exists y \in S . i_R(y) = x$$
$$\Leftrightarrow \text{False} \vee \exists y \in S . i_R(y) = x$$
$$\Leftrightarrow \exists y \in S . i_R(y) = x$$

Since i_R is an injection, it is one–one and so we can pair off the members of $\varnothing \oplus S$ and S. Hence $\varnothing \oplus S \cong S$. The other part, $S \oplus 0 \cong S$, follows by a similar argument.

(1) is proved as follows

The members of $S1$ are in one–one correspondence with some members of $S1 \oplus S2$. The same applies to members of $S2$. However $i_L(S1)$ and $i_R(S2)$ are disjoint by hypothesis, so the sets of members of $S1 \oplus S2$ which are in one–one correspondence with $S1$, $S2$ respectively are disjoint. Furthermore these are the only members of $S1 \oplus S2$, by definition of $\oplus$. The result follows. $\square$

Relation to Programming Languages

The usual algorithmic computer languages implement the arithmetic operators on arithmetic quantities, such as $+$, $-$, $*$, $/$. Some of the less widely used languages such as APL and LISP implement set theoretic quantities and the $\oplus$ operation.

Also the usual algorithmic languages allow data and values to be of certain specific types such as integers, real, Boolean. These types are each sets of values: all the integer values, True and False, etc. The more advanced languages (Pascal, ALGOL 68, Simula, Ada, Chill) allow one to construct one's own types using 'type constructors' such as arrays, structures, or strings. These are all implementations in computer languages of set theoretic construction mechanisms. In particular ALGOL 68 allows a new type to be constructed from two (or more) existing types by means of the union construction, which is precisely an implementation of the set theoretic Disjoint Union which we have just encountered.

Another aspect of the more modern languages (ALGOL 68, Pascal, Ada, Chill) is that a user may define a type which is an explicit set definition such as

$$Colour = \{Red, Green, Yellow, Orange, Blue, Purple, Brown\}$$

and may provide the means of defining a set implicitly by a function procedure which delivers 'True' if the input parameter fulfils the membership criterion and 'False' otherwise.

Thus, behind the scenes as it were, many of the constructs of the programming languages familiar to us are in fact implementations of set theoretic constructs and operations which, in turn, may be used to describe essentials of the problem which is to be addressed.

Example 5.6: Telecommunications

In our telephone exchange example, recall that

$$S \subseteq C^* \times B^* \times TN$$

where S was a set of subscribers, C a set of characters for names, and B a set of characters for addresses. TN was defined

$$TN = D^4 \text{ (quartuples of digits for the telephone number)}$$

We might divide the subscribers into existing subscribers ES and new subscribers NS

$$S = ES \oplus NS$$

where

$$ES \cap NS = \emptyset$$

Example 5.7: Car Supply Database

Again, in our example of cars, the colours could be distinguished between old colours OC and next year's colours NC

$$C = OC \oplus NC$$

where again

$$OC \cap NC = \varnothing$$

Example 5.8: Words over an Alphabet

We can express words of up to three letters over an alphabet A as

$$A^0 \oplus A^1 \oplus A^2 \oplus A^3$$

Theorem 5.3: Distributive Law

The distributive law applies similarly as in arithmetic between $+$ and $\times$

$$(A \times B) \oplus (A \times C) \cong A \times (B \oplus C)$$

Proof
$A \times B \oplus A \times C$ consists of disjoint copies of $A \times B$ and $A \times C$. $B \oplus C$ is built from disjoint copies of B and C. So $A \times (B \oplus C)$ can be divided into $A \times \text{copy\_of\_}B$ and $A \times \text{copy\_of\_}C$. The isomorphism can be defined as follows

$$h: A \times B \oplus A \times C \to A \times (B \oplus C)$$

with

$$h(i_L(a, b)) = (a, i_L(b))$$
$$h(i_R(a, c)) = (a, i_R(c))$$

where $a \in A, b \in B, c \in C$. $\quad\square$

Notation: Distributed Sum

For a finite index set I

$$\bigoplus_{i \in I} A_i$$

or synonymously

$$\sum_{i \in I} A_i$$

are the disjoint unions of the family of sets A_i, $i \in I$; that is

$$\{\text{copy\_of\_}x | \exists i \in I . x \in A_i\}$$

Exercise 5.2

Identifiers in a certain programming language are constructed from a set of letters $A = \{A, B \ldots Z, a, b, \ldots, z\}$ and a set of digits $D = \{0, 1, \ldots, 9\}$. They can be up to six characters in length but must start with a letter. Express the set of possible identifiers using A, D, $\oplus$, $\times$. Are there alternative ways of expressing this? Can you extend it to the set of identifiers of any length?

5.3 HOMOGENEOUS AND HETEROGENEOUS TUPLES AND PRODUCTS

We can distinguish homogeneous tuples, in which the sets of which the components are members are all the same, from heterogeneous tuples, in which they are different.

Definition 5.5: Heterogeneous and Homogeneous Tuples and Products
If a tuple -

$$t \in \underset{i \in I}{\times} A_i$$

is such that

$$A_i = A_j$$

for each $i, j \in I$, the tuple and the Cartesian product to which it belongs are said to be *homogeneous*, in which case we may write

$$\underset{i \in I}{\times} A$$

where $A = A_i$ for all i. Otherwise they are called *heterogeneous*.

Example 5.9

　　$A^n, A^*, A \times A$ are homogeneous

　　$A \times B, \underset{i \in F}{\times} A_i$ are heterogeneous

From theorem 5.1 it follows that

$$\text{Card}\,(A^n) = (\text{Card}\,A)^n$$

and in particular

$$\text{Card}\,(2^n) = (\text{Card}\,2)^n = 2^n$$

5.4 REPRESENTATIONS OF TUPLES IN COMPUTER LANGUAGES

Definition 5.6
We say that a construct in a computer language (or any other language) *represents* a set S of mathematical quantities if, for each member of S, there is a distinct value of the language construct. We may also use the term when there is a distinct value of the language construct for each member of some subset of S.

Example 5.10

(i) The program language data type 'integer' represents $\mathbb{Z}$, the mathematical integers.

Note that in fact some subset of the integers will usually be represented on account of the finite range of language 'integers' (resulting from the computer's word size).

(ii) Tuples may be modelled by many different constructs in programming languages. Arrays are a clear example of such a construct. An array of n integers, for example, can represent a member of

$$\prod_{i=1}^{n} \mathbb{Z}$$

or

$$\mathbb{Z}^n$$

where $\mathbb{Z}$ is the set of integers. An $m \times n$ array of integers may likewise take values in

$$\prod_{1}^{m} \prod_{1}^{n} \mathbb{Z}$$

or

$$(\mathbb{Z}^n)^m$$

or

$$\mathbb{Z}^{m*n}$$

Arrays may hence be used to model homogeneous tuples.

(iii) Fixed-length tuples of heterogeneous sets may be modelled by Structures or Records. For example, if the modes or types A, B, C, D are defined in a program, the construct

$x \, struct(\text{a } A, \text{b } B, \text{c } C, \text{d } D)$

declares a variable x which can adopt values in $A \times B \times C \times D$ where

A, B etc. are the sets of possible values which can be taken by modes A, B etc. respectively.

(iv) Records, as occur in files, normally contain a fixed arrangement of fields of different kinds, and represent tuples in much the same way as structures or records in programming languages. Thus a record in a file in general represents a member of the set.

$$\underset{i=1}{\overset{n}{\times}} A_i$$

where n is the number of components in the record and each A_i is the set of values which can be taken by the mode of the ith component of the record.

A file consists of a sequence of records: the sequence may be of arbitrary length, including zero. Thus a file represents a value of

$$\left(\underset{i=1}{\overset{n}{\times}} A_i \right)^*$$

However most files have a special header record and trailer record. These represent members of

$$\underset{i=1}{\overset{h}{\times}} H_i$$

and

$$\underset{i=1}{\overset{t}{\times}} T_i$$

respectively. Hence the whole file is a value of the set

$$\underset{i=1}{\overset{h}{\times}} H_i \times \left(\underset{i=1}{\overset{n}{\times}} A_i \right)^* \times \underset{i=1}{\overset{t}{\times}} T_i$$

We can show this more clearly by splitting up the definitions

$$File = Head \times Body \times Tail$$

—the set of all files

$$Head = \underset{i=1}{\overset{h}{\times}} H_i$$

—the set of all headers

$$Body = Record^*$$

—the set of all bodies

$$Tail = \underset{i=1}{\overset{t}{\times}} T_i$$

—the set of all trailers

$$Record = \overset{n}{\underset{i=1}{\times}} A_i$$

—the set of all records of a specific type

(v) Strings of various kinds represent values of homogeneous tuples, as we have seen with identifiers. Hence if C is the set of characters, B the set of binary digits, then C^* is the set of character strings, and B^* the set of binary strings. There is a clear connection between B^* and the natural numbers, and for that matter between X^* and the natural numbers where X is any finite set: members of X^* are numerals in the base $Card(X)$ which represent a natural number.

Exercise 5.3

Numbers in a certain programming language are expressed in any one of the following ways

(i) A non-empty string of digits, possibly preceded by a '+' or a '−'.
(ii) As above with a decimal point at any position after the leading sign.
(iii) As (ii) above but followed by 'E' and a string of digits, possibly preceded by a '+' or a '−'.

Express the set of all such number representations using the notation, $\oplus$, $\times$, $*$ learned in this chapter. Singleton sets may be defined

$$Exp = \{'E'\}$$

etc. as required.

5.5 STRINGS, SEQUENCES, LISTS

Definition 5.7: Strings
A *string*, *sequence* or *list* over a set A is the set of all tuples A^*. This is a useful concept: we can define various operations on strings. First, concatenation.

Definition 5.8: Concatenation

$$\| : A^* \times A^* \rightarrow A^*$$

The operation $\|$ takes two strings as its argument and produces a string as its result. The two argument strings are joined to make a new string

$$\langle a_1, a_2, \ldots, a_n \rangle \| \langle b_1, b_2, \ldots, b_m \rangle \triangleq$$
$$\langle a_1, a_2, \ldots, a_n, b_1, b_2, \ldots, b_m \rangle$$

It follows from the definition that $\parallel$ is associative—that is, given any three tuples $S1, S2, S3$, $S1\parallel(S2\parallel S3)=(S1\parallel S2)\parallel S3$ and so one may write without ambiguity $S1\parallel S2\parallel S3$.

Another operator, hd, pronounced 'head', yields the first element of the string.

Definition 5.9

$$\text{hd}: A^* \to A$$
$$\text{hd}(\langle a_1, \ldots, a_n \rangle) \triangleq a_1$$

Note that hd $\langle \ \rangle$ is not defined. Note also that $\text{hd}(\langle a_1, \ldots, a_n \rangle) \neq \langle a_1 \rangle$.

Another operator, tl, pronounced 'tail', returns the string with the first element removed.

Definition 5.10

$$\text{tl}: A^* \to A^*$$
$$\text{tl}(\langle a_1, \ldots, a_n \rangle) \triangleq \langle a_2, \ldots, a_n \rangle$$

Note that tl $\langle \ \rangle$ is not defined.

Theorem 5.4

If $x \in A^*$ and $x \neq \langle \ \rangle$

$$\langle \text{hd } x \rangle \parallel \text{tl } x = x$$

Proof
The proof is left to the reader.

Discussion

Thus strings are therefore just lists of things of the same type. As such, they are rather like the familiar Stacks, especially when taken with the hd and tl operators. hd presents the top-most element, and tl removes the top-most element, so that these two together act like the familiar POP operation. PUSH, the act of adding a new element on to a stack, is achieved by the function $f(x, s) \triangleq \langle x \rangle \parallel s$.

Hence there is no essential difference between strings and stacks.
Further operations can be defined.

Definition 5.11: Length

$$\text{len}: A^* \to \mathbb{N}$$

gives the number of elements in a string. Len can be defined as the operation satisfying

$$\text{len} \langle\ \rangle = 0$$
$$\text{len } S = \text{len tl } S + 1$$

Theorem 5.5

$$\text{len } (S1 \| S2) = \text{len } S1 + \text{len } S2$$

Proof
We prove by induction on the length of *S1*.

(i) If

$$\text{len } S1 = 0$$

then

$$S1 = \langle\ \rangle$$

in which case

$$\text{len } S1 \| S2 = \text{len } S2$$
$$= \text{len } S1 + \text{len } S2$$

(ii) We assume the theorem is true for len $S1 = n$, and show it is true for len $S1 = n + 1$. Let len $S1 = n + 1$ and let $S1 = \langle x \rangle \| S0$. Then

$$\text{len } (S1 \| S2) = \text{len } ((\langle x \rangle \| S0) \| S2)$$
$$= \text{len } (\langle x \rangle \| (S0 \| S2))$$

by associativity of $\|$

$$= 1 + \text{len } (S0 \| S2)$$

Now len $(S0) = n$ and by our induction hypothesis

$$= 1 + \text{len } (S0) + \text{len } (S2)$$
$$= \text{len } (S1) + \text{len } (S2)$$

$\square$

Brackets ( ) are used for indexing: $S(n)$ gives the *n*th element of *S*.

Definition 5.12: Indexing

$$\_(\_): A^* \times \mathbb{N} \to A$$

The _s indicate where, syntactically, one places the two arguments.

Indexing $S(n)$ satisfies

$S(1) = $ hd S

$S(n+1) = ($tl $S)(n)$

$S(n)$ is defined only for $n < $ len S

Theorem 5.6

$(S1 \| S2)(n) = $ if $n \leqslant $ len $S1$ then $S1(n)$

else $S2(n - $ len $S1)$

Proof
We prove this induction on the length of $S1$.

(i)   Let len $(S1) = 1$ and $S1 = \langle x \rangle$. Then

$(S1 \| S2)(n) = (\langle x \rangle \| S2)(n)$

$= x$ if $n = 1$

$= S2(n-1)$ if $n > 1$

Since $n \in \mathbb{N}$ we cannot have $n < 1$ and so the formula is proved for len $S1 = 1$.

(ii) We now show that if the theorem holds for len $S1 = k$, then it holds for len $S1 = k+1$.
Suppose len $S1 = k+1$ and $S1 = \langle x \rangle \| S0$ where len $S0 = k$. Then

$(S1 \| S2)(n) = (\langle x \rangle \| S0 \| S2)(n)$

$= x$ if $n = 1$

$= (S0 \| S2)(n-1)$ if $n > 1$

However len $S0 = k$ so by the inductive hypothesis

$(S0 \| S2)(n-1) = S0(n-1) = S1(n)$ if $n - 1 < $ len $S0$

$= S2(n - 1 - $ len $S0)$

$S2(n - ($len $S0 + 1))$

$= S2(n - $ len $S1)$ otherwise   □

Many other useful string operators can be defined.

Definitions 5.13
dels: $A^* \times \mathbb{N} \to A^*$
which is such that dels(S, n) produces a new string by removing the nth element of string S

mods: $A^* \times \mathbb{N} \times A \to A^*$
which is such that mods(S, n, e) replaces the nth element of S with e

subs: $A \times \mathbb{N} \times \mathbb{N}_0 \rightarrow A^*$

which is such that subs (S, n, k) produces the sub-string of S consisting of elements $\langle S(n), S(n+1), \ldots, S(n+k) \rangle$

Many others can be thought up.

Exercise 5.4

What are the following

(i)   hd tl $\langle a, b, c \rangle$
(ii)   len $\langle \langle a, b \rangle, \langle b, c \rangle, \langle c, d \rangle \rangle$
(iii)   len $\langle a, a, a \rangle$
(iv)   tl hd $\langle \langle a, b \rangle, \langle b, a \rangle \rangle$
(v)   hd $(\langle a, b, c \rangle \| \langle c, b, a \rangle)$
(vi)   tl $(\langle a, b, c \rangle \| \langle c, b, a \rangle)$
(vii)   tl $(\langle a \rangle \| \langle b, c \rangle)$
(viii) tl $(\langle a, b, c \rangle \| \langle d, e, f \rangle)(2)$
(ix)   tl $(\langle a, b, c \rangle \| \langle d, e, f \rangle)(3)$

Exercise 5.5

What observations can you make about the operators dels, mods, subs relating them to the other operators and to each other? For example, the relationship between len S and len dels (S, n) etc.

Exercise 5.6 (Workshop)

If a workshop exercise is required, review your current workshop data definitions to see if more complete or convenient formulations can be achieved using tuples etc. learned in this chapter.

SUMMARY OF RESULTS

Result	*Reference*
For finite sets	Theorem 5.1
$\quad$ Card $(A \times B) = $ Card $A *$ Card B	
$\quad$ Card $\underset{i \in F}{\times} A_i = \underset{i \in F}{\prod}$ Card A_i	
Properties of $\oplus$	Theorem 5.2
	(continued overleaf)

Summary of Results (continued)

Result	Reference
$(A \times B) \oplus (A \times B) \cong A \times (B \oplus C)$	Theorem 5.3
$\langle \text{hd } x \rangle \| \text{tl } x = x$	Theorem 5.4
$\text{len}(S1 \| S2) = \text{len } S1 + \text{len } S2$	Theorem 5.5

Chapter 6 Mappings and Functions

NOTATION INTRODUCED

Notation	Concept	Reference
$f: A \to B$ $a \overset{f}{\mapsto} b$	function	Def. 6.1
	codomain	Def. 6.1
	total, partial functions	Def. 6.2
	domain, range	Def. 6.3
Graph(f)	graph of a function	Def. 6.4
	image	Def. 6.5
$A \to B$ A^B		Def. 6.6
	onto, into, surjection	Def. 6.7
	injection, one–one	Def. 6.8
	bijection	Def. 6.9
	same cardinality	Def. 6.10
	finite, infinite, countable, uncountable	Def. 6.11
M^{-1}	inverse mapping	Def. 6.12
	representation functions	Def. 6.13
$g \circ f$	composition	Def. 6.14
	tables	Def. 6.15
	algorithms, procedure	Def. 6.16
	pre-condition	Def. 6.17

Sets alone are of limited use. We need ways of associating sets together. This can be done by means of mappings, functions and relations. The terms 'function' and 'mapping' are usually used synonymously by mathematicians to denote the same concept. When one represents this concept in programming languages, one normally finds that a distinction arises between two kinds of function or mapping. One kind is fixed and usually not finite, while the other is 'variable' or 'dynamic' and usually finite. The first kind is typically 'represented', that is implemented, as a program function or procedure, whereas the second kind is typically implemented by means of a variety of program language data structures such as arrays.

6.1 NOTATION

Definition 6.1: Functions and Mappings
A *function* or *mapping* f from a set A to a set B is a 'rule' or 'method' which pairs elements of the set A with unique elements of the set B. We write

$$f:A \rightarrow B$$

to indicate that f is a mapping from the set A to the set B. B is called the *codomain* of the mapping f. (A is sometimes called its *domain*. We shall avoid this usage as it conflicts with another use of the term; see definition 6.2.)

For each element a of A, f defines an element b of B. We write

$$a \overset{f}{\mapsto} b \qquad a \in A, b \in B$$

or

$$a \overset{f}{\mapsto} f(a)$$

Example 6.1

Consider the population expressed by the set

$$World = \{x | \text{Is-human}(x)\}$$

and various attributes

$$Years = \{n | n \in \mathbb{N}_0 \wedge 0 \leqslant n \leqslant 120\}$$
$$Colour = \{\text{Black, Brown, Blonde}\}$$
$$Gender = \{\text{Male, Female}\}$$

Each member of the population has one attribute out of each set which represents an attribute, so that each has a specific age, each a gender, etc. This possession of attributes can best be represented by a mapping, which associates a member of the codomain for each member of the domain. We can thus define a mapping called *Age*

$Age: World \rightarrow Years$

indicating that the mapping *Age* gives a member of the set *Years* for each member of the set *World*: everyone in the *World* has an *Age* which is in the set $0 \ldots 120$. This can be expressed

$$\forall x \in World . \exists n \in Years . x \overset{Age}{\longmapsto} n$$

or equivalently

$$\forall x \in World . \exists n \in Years . Age(x) = n$$

Likewise for the other attributes

$Sex: World \rightarrow Gender$

$Hair\text{-}colour: World \rightarrow Colour$

Definition 6.2: Total/Partial Functions
A function $f: A \rightarrow B$ is called *total* if the domain is the total set A, and otherwise is called *partial*.

Thus in example 6.1 Age and Sex are total functions because every member of World will be associated with some member of Years and Gender. On the other hand the function Hair-colour is partial because (presumably) some members of World will have hair colours not in the set {Brown, Black, Blonde}, say Grey or Bald for example. When we take a function f, therefore, we have the following definitions.

Definition 6.3: Domain, Range
Consider

$f: A \rightarrow B$

The *domain* of f is that subset of A for which f gives a value in B, and the *range* is that subset of B of elements to which f maps a value. Thus the range of f is the set

$$\mathrm{rng}(f) = \{b | b \in B \wedge \exists a \in A . f(a) = b\}$$

and the domain of f is the set

$$\mathrm{dom}(f) = \{a | a \in A \wedge \exists b \in B . f(a) = b\}$$

The same terms are used with the same meanings in the context of functions of finite domains (mappings).

A pictorial representation of a mapping is given in figure 6.1.

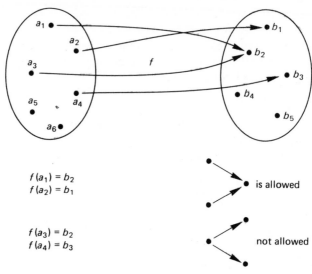

$$f(a_1) = b_2$$
$$f(a_2) = b_1$$

$$f(a_3) = b_2$$
$$f(a_4) = b_3$$

is allowed

not allowed

Figure 6.1

Example 6.2

In school mathematics one learns about many functions with domain and range consisting of the Real numbers.
    For example

    Times-2: $\mathbb{R} \to \mathbb{R}$

      $x \mapsto 2 * x$

    $(x \mapsto \text{Times-2}(x))$

    Square: $\mathbb{R} \to \mathbb{R}$

      $x \mapsto x^2$

    $(x \mapsto \text{Square}(x))$

    Sin: $\mathbb{R} \to \mathbb{R}$

      $x \mapsto \text{Sin}(x)$

One may picture such functions as diagrams in which an arrow is drawn from each point of the domain to the corresponding point of the range. See figure 6.2.

6.2 GRAPHS

A familiar way of depicting a function is by means of its graph. In its familiar form a graph of a function $f: A \to B$ is depicted as a curve on a

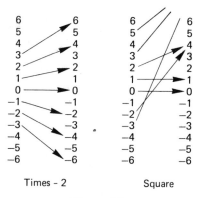

Times - 2   Square

Figure 6.2

plane such that the two co-ordinates of any point on the curve give a pair of values $x \in A$, $y \in B$ where

$$x \overset{f}{\mapsto} y$$

that is

$$y = f(x)$$

Example 6.3

The pictorial graphs in figures 6.3–6.5 are simply reconstructions of all the pairs of values (x, y) such that

$$x \overset{f}{\mapsto} y$$

Definition 6.4

A *graph* of a function is a set of pairs, so it is a subset of the Cartesian product $A \times B$ where $f: A \to B$. For $f: A \to B$ we define

$$\text{Graph}(f) = \{(a, b) \mid a \in A \wedge b \in B \wedge b = f(a)\}$$

Thus another way of looking at a function or a mapping is to consider it as a set of pairs (that is, its graph): a mapping $M: A \to B$ is such that $M \subseteq A \times B$ where

$$\forall x \in A . \forall y, z \in B . (x, y) \in M \wedge (x, z) \in M . \Rightarrow y = z$$

which expresses the 'uniqueness property' that each member of the domain of a map is mapped to a unique member of its range.

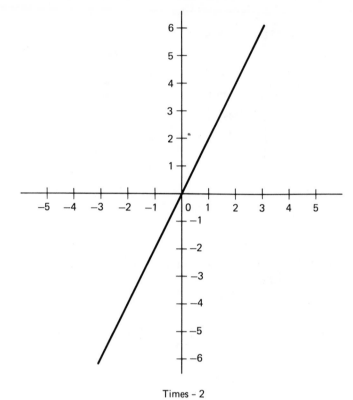

Times – 2

Figure 6.3

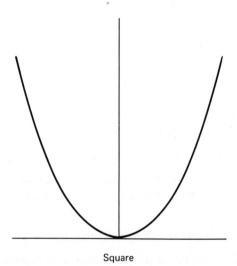

Square

Figure 6.4

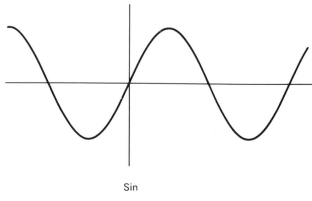

Sin

Figure 6.5

Example 6.4: Character Sets

Here is a more computer-oriented example. The ASCII code represents each member of the ASCII character set as a sequence of eight binary bits. Let

$$ASCII = \{NUL, SOH, \ldots\}$$
$$B = \{0, 1\}$$

Then we have the mapping

$$ASCII\text{-}code: ASCII \to B^8$$

normally represented by a table of pairs (another way of representing a graph) as follows

NUL	00000000
SOH	00000001
⋮	⋮
Space	00100000
⋮	⋮
0	00110000
1	00110000
⋮	⋮

6.3 PROPERTIES OF MAPPINGS AND FUNCTIONS

There are various terms which describe different sorts of function. From the definition of graphs, it follows that a function $M: A \to B$ is total (definition 6.2) if for each $a \in A$ there is a pair $\langle a, b \rangle \in \text{Graph}(M)$

$$\forall a \in A \,.\, \exists b \in B \,.\, \langle a, b \rangle \in \text{Graph}(M)$$

and if the negation of this condition is true, the function is partial.

Definition 6.5: Images
If $M(a) = b$, b is called the *image* of a under M.

Definition 6.6
$$A \rightarrow B \text{ or } B^A$$

denotes the set of all the possible mappings from a domain in A to a range in B. Thus we could write

$$M : A \rightarrow B$$

The ':' is pure convention; it is the same as '$\in$'.
    We call '$A \rightarrow B$' the 'type' of the mapping M.

Discussion
In general if we have some symbol x representing a value, it is a value of some type. Thus if we write an expression

$$x + 2$$

this has little meaning until we say what is the type of x

$$x \in \mathbb{Z}$$

or

$$x : \mathbb{Z}$$

indicating that its type is integer. This is precisely the same as saying that x belongs to the set of integers. Likewise, when we write

$$M \in A \rightarrow B$$

or

$$M : A \rightarrow B$$

we mean that M is of type 'mapping from a domain in A to a range in B' or M is a member of $A \rightarrow B$.

Example 6.5

(i) $x + 2$ where $x \in \mathbb{Z}$ maps $\mathbb{Z} \rightarrow \mathbb{Z}$
(ii) $\sqrt[+]{x}$ where $x \in \mathbb{R}$ maps $\mathbb{R} \rightarrow \mathbb{C}$ ($\mathbb{C}$ = complex numbers)
(iii) $x/x + 1$ where $x \in \mathbb{Z}$ maps $\mathbb{Z} \rightarrow \mathbb{R}$

(iv) $x \text{ DIV } y$ where $x, y \in \mathbb{Z}$ maps $\mathbb{Z} \times \mathbb{Z} \to \mathbb{B}$

   (DIV gives TRUE if y is a divisor of x, FALSE otherwise)

(v) $(a, b, c) \mapsto (c, a, b)$ where $a, b, c \in S$ maps triples of elements of S into permuted triples of elements of S, so is of type

$$S^3 \to S^3$$

Example 6.6: Personnel File

In our personnel file example we said that the set of employees could be characterised by a Cartesian product

   Names × Ages × Grades × Salaries

However this does not capture the fact that each employee is associated with just one *Age*, *Grade* and *Salary*. For example, the following two entries could co-exist in the personnel file as it stands

   ⟨"John Smith", 26, 18A, £12 000⟩

   ⟨"John Smith", 26, 19B, £13 000⟩

To constrain the file to containing each employee's name only once, we can use a mapping construction and define *Employee* as

   Employees = Names → (Ages × Grades × Salaries)

Since the file will not contain employees with all possible names, this will be a partial mapping. In some literature partial mappings are indicated by using the $\mapsto$ symbol.

   In fact, of course, one may well have two or more employees with the same name, and in practice one assigns to each employee some unique identifier or number to cater for this eventuality. In this case we would define

   Employees = Employee-Id → (Names × Ages × Grades × Salaries)

Example 6.7: Time-table

In our time-tabling problem, the various constraints, such as no teacher can teach in more than one lesson in each period, and no class can have more than one lesson in each period, can also be conveniently expressed using mappings. Events were defined as

   Events ⊆ P × T × C × L

where *P*, *T*, *C*, *L* were Periods, Teachers, Classes and Lessons respectively. We can express the constraints by two mappings in place of the *Events* construct

$$Teacher\text{-}schedule \in T \rightarrow (P \rightarrow (C \times L))$$
$$Class\text{-}schedule \in C \rightarrow (P \rightarrow (T \times L))$$

and to ensure that these two schedules correspond, we must also have

$$\forall t \in T . \forall p \in P . \forall c \in C . \forall l \in L .$$
$$(Teacher\text{-}schedule(t)(p) = \langle c, l \rangle \Leftrightarrow$$
$$Class\text{-}schedule(c)(p) = \langle t, l \rangle)$$

Let us examine these statements. The *Teacher-schedule* is a mapping from each teacher to another mapping: that is, each teacher has his/her own personal schedule. This schedule consists of a mapping from periods to class–lesson pairs, so for each period the teacher has a particular class for a particular lesson.

Likewise the *Class-schedule* is a mapping from each class to another mapping, which is in turn a schedule consisting of a mapping from periods to teacher–lesson pairs.

The constraint says that if the *Teacher-schedule* allocates class c and lesson l during period p to teacher t, then the *Class-schedule* must allocate teacher t and lesson l to class c during period P, and vice versa. Such constraints which must invariably hold on a system, are often called 'invariants'.

Exercise 6.1

Recall that we previously (chapter 4) included *Rooms* in the *Lessons* and *Events* so that

$$Events \subseteq P \times T \times C \times L \times Rooms$$

How can we restate this in terms of mappings so as to show that only one lesson may take place in each room at once?

Exercise 6.2

Which mappings in all those mentioned become partial if

(i)   Teachers may have free periods?
(ii) Classes may have free periods?
(iii) Rooms may sometimes be empty?

6.4 MORE PROPERTIES OF MAPPINGS

Mappings may have various characteristics.

Definition 6.7: 'onto', 'into'
A mapping

$$M:A \rightarrow B$$

is said to be from *A onto B* or a *surjection* if for every $b \in B$ there is an $a \in A$ such that $M(a) = b$, in other words if the range of the mapping is the whole of *B*. Formally

$$\forall b \in B . \exists a \in A . M(a) = b$$

For general mappings which are not necessarily 'onto', one speaks of $M:A \rightarrow B$ mapping *A into B*.

Definition 6.8: Injections
A mapping $M:A \rightarrow B$ is *one–one* or an *injection* if for each $b \in B$ there is at most one $a \in A$ such that $M(a) = b$. Formally

$$\forall a, c \in A . M(a) = M(c) \Rightarrow a = c$$

Thus in an injection we never have the situation illustrated in figure 6.6.

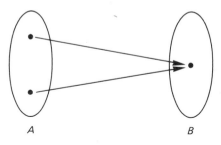

Figure 6.6

Definition 6.9: Bijections
A total, one–one mapping from *A* onto *B* is called a *bijection*. In this case there is a one-to-one pairing or 'correspondence' of the members of *A* and *B*. They are sometimes called 'isomorphic' which means 'having the same shape'. It is clear that in the case of finite sets, *A* and *B* must have the same cardinality. Thus a set *A* has cardinality *n* if and only if there is a bijection from *A* to $[1 \ldots n]$, the set of integers between 1 and *n* inclusive.
We can now give a new definition of cardinality.

Definition 6.10: Cardinality
Two sets A and B have the same *cardinality* if there is a bijection from A to B. A set A is of *cardinality* n if there is a bijection from A onto $\{1 \ldots n\}$.

Example 6.8

(i)   In figure 6.7 (a, b, c) has cardinality 3.

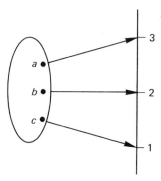

Figure 6.7

(ii) The set of integers has the same cardinality as the set of even integers (figure 6.8).

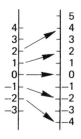

Figure 6.8

There is a bijection from the integers to the even integers given by, for example, $M: x \mapsto 2 * x$.

Bijections can be used to define sets whose cardinality is: finite, infinite, countable, uncountable, where these are defined as below.

Definition 6.11
A set A is *finite* if there is a bijection from A to a finite subset of the

natural numbers which is definable as

$$\{x | x \in \mathbb{N} \land x \leqslant n\}$$

for some natural number n.

A set A is *infinite* if it is not finite.

A set A is *countable* if there is a bijection from A to some subset of the natural numbers.

A set A is *uncountable* if it is not finite and not countable.

Examples 6.9

Some examples of various kinds of sets are illustrated by the following theorems.

Theorem 6.1

The set of integers $\mathbb{Z}$ is countable.

Proof
There is a bijection

$$f : \mathbb{Z} \to \mathbb{N}$$
$$f(x) \triangleq -2*x-1 \qquad \text{if } x < 0$$
$$2*x \qquad \text{if } x \geqslant 0$$

Theorem 6.2

The even natural numbers are countable.

Proof
The even natural numbers comprise the set
$$\{2*x | x \in \mathbb{N}\}$$

Thus there is a bijection

$$f : \{2*x | x \in \mathbb{N}\} \to \mathbb{N}$$
$$f(x) \triangleq x \div 2$$

Theorem 6.3

The set of all pairs of natural numbers is countable.

Proof
We consider a mapping which carried out a 'diagonalisation' process

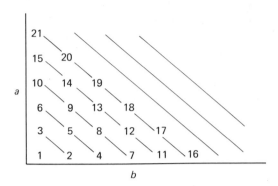

Figure 6.9

illustrated in figure 6.9. The bijection

$$f:(\mathbb{N} \times \mathbb{N}) \to \mathbb{N}$$
$$f(a, b) \triangleq (a + b) * (a + b - 1)/2 - b + 1$$

carries each pair of natural numbers to a unique image which is a natural number.

Corollary
Hence $\mathbb{Q}$, the set of rational numbers, is countable (since all rational numbers can be expressed as a pair of integers a/b and the integers are countable by theorem 6.1).

Theorem 6.4

The real numbers are countable.

Proof
We use a *reductio ad absurdum* argument. We assume that the real numbers are countable and infer a contradiction. First a lemma.

Lemma
If a set S is countable then any subset $T \subseteq S$ is countable. (The proof is deferred for the moment.)

Fom the lemma it follows that if a subset of S is not countable then S is also not countable. We shall prove that the subset of the reals comprising the numbers in the interval

$$0 \leqslant x < 1$$

is not countable, and from this and the lemma it will follow that the reals are countable.

All the numbers in the range

$$0 \leqslant x < 1$$

are representable by an expansion in some base (say 10 because it is more familiar), and furthermore each expansion of the form below represents a number in that range

$$. d_1 d_2 d_3, \ldots,$$

expanded indefinitely.
  (There is some ambiguity, for example

$$. 5000 \ldots$$

and

$$. 4999 \ldots$$

represent the same number. However, we can avoid the consequences of this.)
  If these numbers are countable, they can be ordered so that each number has a unique index n. Let the expansion of the ith number be

$$. _id_1 \, _id_2 \, _id_3, \ldots, \, _id_n, \ldots,$$

(where a number has two possible expansions, the choice is arbitrarily made). We construct a new number in whose expansion the ith digit is given by

$$\text{if } _id_i = 0 \text{ or } _id_i = 1 \text{ then 2 else 1}$$

This new expansion differs from the ith expansion in our ordered set in the ith digit. Furthermore it cannot be an alternative representation of a number already represented because it contains no 0 or 9. Therefore it represents a number not in our ordered set, contradicting the assumption that we can enumerate all the numbers between 0 and 1. Consequently, from the lemma, the reals cannot be enumerated. □

Proof of Lemma
Let S be a countable set and $T \subseteq S$. Then we can give a unique index to each member of S, denoted

$$\text{ind}: S \rightarrow \mathbb{N}$$
$$\text{ind}(x)$$

Consider the set of natural numbers

$$P \triangleq \{n | n \in \mathbb{N} \land \exists x \in T . n = \text{ind}(x)\}$$

We can then define an index on the members of P as follows

$$\text{ind}'(1) \triangleq 1 \text{ if } 1 \in P$$
$$0 \text{ if } 1 \notin P$$
$$\text{ind}'(n+1) \triangleq \text{ind}'(n) \quad \text{if } n+1 \notin P$$
$$\text{ind}'(n)+1 \text{ if } n+1 \in P$$

ind' restricted to P is the index we require. It is one–one because if $x < y$ and $y \in P$ we must have $\text{ind}'(x) < \text{ind}'(y)$. ind is one–one on S from the assumption that S is countable, and therefore $\text{ind}'(\text{ind}(x))$ for $x \in T$ maps T to $\mathbb{N}$, and so T is countable.   □

Theorem 6.5

The set of all mappings $\mathbb{N} \to \mathbb{N}$ is uncountable.

Proof
Again, using *reductio ad absurdum*, assume the mappings

$$\mathbb{N} \to \mathbb{N}$$

are countable, order them accordingly and let m_i be the ith mapping in the ordering. We can then construct a mapping which is different from the m_i as follows

$$m'(i) \triangleq 1 + m_i(i)$$
$$\text{for all } i \in \mathbb{N}$$

Thus for all $i \in \mathbb{N}$, $m'(i) \neq m_i(i)$ and therefore $m' \neq m_i$, contradicting the assumption that all $m \in \mathbb{N} \to \mathbb{N}$ can be counted and indexed.

Theorem 6.6

$\mathscr{P}(\mathbb{N})$ is uncountable.

Proof
Again by *reductio ad absurdum*, assume $\mathscr{P}(\mathbb{N})$ is countable and enumerate its members S_i, $i \in \mathbb{N}$. We construct a new subset not the same as any S_i

$$S \triangleq \{x \,|\, x \in \mathbb{N} \wedge x \notin S_x\}$$

which contains each $x \in \mathbb{N}$ if and only if $x \notin S_x$. S must therefore be different from every S_x, contradicting the assumption.

Definition 6.12: Inverse Mappings
If a mapping $M : A \to B$ is one–one, then M has an *inverse mapping*

$$M^{-1} : B \to A$$

such that

$$M^{-1}(b) = a \Leftrightarrow M(a) = b$$

Intuitively one can see that if a mapping is one–tone, one can reverse the direction of the maps. See figure 6.10.

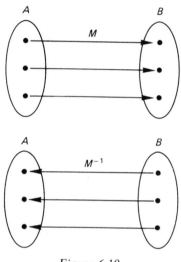

Figure 6.10

Theorem 6.7

$$(M^{-1})^{-1} = M$$

Proof

$$(M^{-1})^{-1}(b) = a \Leftrightarrow M^{-1}(a) = b$$
$$\Leftrightarrow M(b) = a$$

Hence

$$(M^{-1})^{-1} = M$$

Definition 6.13: Representation of Sets
A function or mapping is said to *represent* a set A in a set B if it maps $A \to B$ one to one.

Example 6.10: Character Sets

The ASCII code example we saw earlier represents elements of the ASCII character set by a string of eight binary bits

$$Code: ASCII \rightarrow B^8$$

a one–one mapping which is usually defined by a table

NUL	$\mapsto$	0000 0000
SOH	$\mapsto$	0000 0001
$\vdots$		$\vdots$
space	$\mapsto$	0010 0000
$\vdots$	$\vdots$	$\vdots$
0	$\mapsto$	0011 0000
1	$\mapsto$	0011 0001
$\vdots$	$\vdots$	$\vdots$

Example 6.11: Representations in Programming Languages

Other data types which can be used or defined in a programming language are represented by a compiler as some kind of internal form which is distinguishable at run time by the semantics, that is by the interpretation expressed by the execution. If the compiler is correct therefore, the representation of TRUE will be distinguished from that of FALSE, the representation of 400 from that of 399, the representation of the member of a set $\{a, b, d, c, f\}$ from that of e, and so on. In general, the total data domains of a program, which can be considered as the Cartesian product of the types of (that is, the values attainable by) all the variables on the program, is represented by the compiler as another internal domain of binary patterns. Any particular value of this domain is the state of the program, and is represented by a unique internal state which is interpreted appropriately by the object program at run time.

When casting a problem in terms of a programming language, the programmer represents the aspects of the problem, including the information content, as a collection of language constructs, which include the data. We shall see in later examples how various data domains can be variously represented by language facilities.

Recapitulation

A mapping is the same as a function, except that in a lot of literature a mapping is taken to be finite, that is the domain and range are finite sets. We are all accustomed to functions of numeric quantities such as $2*x$, $x**2$, $x+3$ etc. Thus one would write

$$f(x) = x^2$$
$$f(x) = x + 3$$

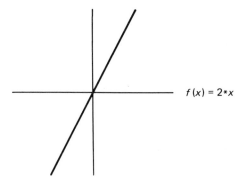

Figure 6.11

$f(x) = 2*x$

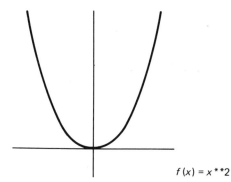

Figure 6.12

$f(x) = x**2$

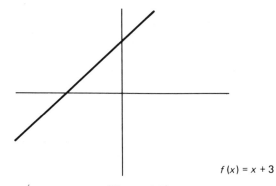

$f(x) = x + 3$

Figure 6.13

etc. For these functions we are used to the idea of drawing graphs. See, for example, figures 6.11 to 6.13.

Any function $f:A \rightarrow B$ has a graph consisting of all the pairs

$$\{(a, b) | a \in A \wedge b \in B \wedge f(a) = b\}$$

Examples 6.12 (from Algebra)

In algebra the operations associate tuples of values with a unique further value: for example, addition in arithmetic associates every pair of numbers with a unique third number, namely their sum

$$+:\mathbb{R} \times \mathbb{R} \to \mathbb{R}$$
$$(x, y) \mapsto +(x, y)$$

where $+(x, y)$ the Reverse Polish notation, normally written in the infix form $x + y$, is the sum of x and y. Likewise for other arithmetic operations

$$*:\mathbb{R} \times \mathbb{R} \to \mathbb{R}$$
$$(x, y) \mapsto *(x, y)$$
$$-:\mathbb{R} \times \mathbb{R} \to \mathbb{R}$$
$$(x, y) \mapsto -(x, y)$$
$$-:\mathbb{R} \to \mathbb{R}$$
$$x \mapsto -x$$

Similarly for the Boolean algebra with values TRUE, FALSE

$$\wedge:\mathbb{B} \times \mathbb{B} \to \mathbb{B}$$
$$\vee:\mathbb{B} \times \mathbb{B} \to \mathbb{B}$$
$$\sim:\mathbb{B} \to \mathbb{B}$$

6.5 COMPOSITION OF FUNCTIONS AND MAPPINGS

Definition 6.14: Composition
Let f, g be two functions

$$f:A \to B$$
$$g:B \to C$$

we can combine them to form a third function called their *composition* and written

$$g \circ f:A \to C$$
$$g \circ f(a) \triangleq g(f(a))$$

This is illustrated in figure 6.14.

The relationship between the functions and their domains and ranges is indicated in the arrow diagram in figure 6.15.

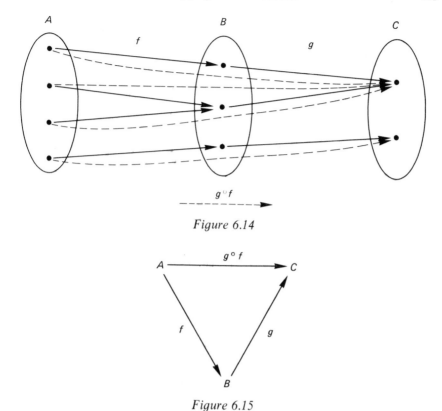

Figure 6.14

Figure 6.15

Example 6.13: Personnel File

Let us elaborate our personnel file example a little further. To recapitulate, in chapter 4 we showed how the file of employees could be represented by a Cartesian product

Employees = Names × Ages × Grades × Salaries

However, this is not entirely satisfactory. It means that any two members of the Cartesian product could represent two distinct employees: thus one could have two employees with the same name, age and grade but different salaries for example. This is unlikely, but perhaps not impossible. We need an 'employee identifier', distinct from the name, to distinguish employees. Suppose we define

Employees = Employee-id × Names × Ages × Grades × Salaries

This is still not quite appropriate because two different employees could be represented by the same *Employee-id* but with different names, and that is not what we want.

The requirement can be captured by the following construct

$$Employees = Employee\text{-}id \rightarrow$$
$$(Names \times Ages \times Grades \times Salaries)$$

or alternatively by four separate mappings

$$M_n = Employee\text{-}id \rightarrow Names$$
$$M_a = Employee\text{-}id \rightarrow Ages$$
$$M_g = Employee\text{-}id \rightarrow Grades$$
$$M_s = Employee\text{-}id \rightarrow Salaries$$

In either of these forms, each *Employee-id* is associated with just one *Name, Age, Grade* and *Salary.*

In place of the fourth-mapping M_s, if all employees of any given grade were paid the same salary, so that the grade determined the salary uniquely, we could instead have the following mappings

$$M_n = Employee\text{-}id \rightarrow Names$$
$$M_a = Employee\text{-}id \rightarrow Ages$$
$$M_g = Employee\text{-}id \rightarrow Grades$$
$$Rate = Grades \rightarrow Salaries$$

Now to find the salary of any employees we apply first the function M_g to the *Employee-id*, and then the function *Rate* to the resulting grades

$$Wage = Rate \circ M_g$$
$$Wage\text{:}Employee\text{-}id \rightarrow Salaries$$

Examples 6.14: Programs

(This section may be omitted on a first reading.)

A program may be considered as representing a function which associates an output value with an input value

$$program\text{:}Inputs \rightarrow Outputs$$

Of course, different programs will accept different kinds of inputs and produce different kinds of outputs. These may be complex types, consisting of many parts—Cartesian products in fact. A program which calculates the mean and standard deviation of a list of numbers for example would represent a function

$$f\text{:}\mathbb{R}^+ \rightarrow \mathbb{R} \times \mathbb{R}$$

(Recall that $\mathbb{R}^+$ is a list of at least one real number.) In fact, the program *represents* the function, so that there is a one–one mapping from the

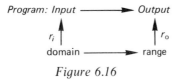

Figure 6.16

function to the program where the domain and range of the functions are mapped one–one to the input and output of the program. See figure 6.16. Then

$$Program^{\circ} r_i = r_o{}^{\circ} f$$

For example, numbers may be represented by strings of digits, lists of numbers may be represented by a list of digit strings followed by a terminating symbol, or by a number-representation giving the length of the list followed by the list of digit-strings, or by some other syntactic form.

We have already shown that the state of a program is the Cartesian product of the types of (values attainable by) all its variables. Thus every program language statement S within a given program represents a mapping from the state to a new state

$$State \rightarrow State$$

If we define a composite statement

$$S = S_1; S_2$$

where the semicolon is the usual program statement separator, and a 'representation function' R

$$R: Statements \rightarrow (State \rightarrow State)$$

then

$$R(S) = R(S_2)(R(S_1))$$

that is

$$R(S) = R(S_2)^{\circ} R(S_1)$$

so that the statement combinator ';' represents functional composition illustrated by the arrow diagram in figure 6.17.

We shall see more about the 'meaning' or semantics of program language statements in chapter 12. One should note that if a program represents a partial function then it does not have a defined effect when applied to a state outside its domain. Of course, good programming practice would decree that an effect from which the user can recover is desirable.

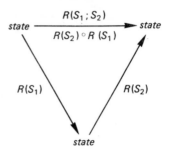

Figure 6.17

6.6 FURTHER EXAMPLES

Example 6.15: Expressions

Expressions provide another example of mappings. Any expression denotes a mapping from the Cartesian product of the types of its variables to the type of its result. Consider for example

$$2x^2 + 3y - z$$

where x, y and z are Real variables. This expression maps

$$\mathbb{R} \times \mathbb{R} \times \mathbb{R} \to \mathbb{R}$$

If we name the expression E, that is

$$E \triangleq 2x^2 + 3y - z$$

then

$$E : \mathbb{R} \times \mathbb{R} \times \mathbb{R} \to \mathbb{R}$$

and

$$E(7.0, 3.1, 2.6) = 104.7$$

We find expressions abounding in programming languages, in the actual parameters of procedure calls, on the right-hand sides of assignment statements, and as the indices of arrays, to name just a few. The types of the variables and the result may be different also. Consider the following expression, which is allowed in ALGOL and various other high level languages

if B **or** C **then** $x + 7.0$ **else** $x - 7.0$

B and C are Boolean variables, and x is Real. The type of the whole expression must be the same as that of the **then** and **else** clauses, which must be the same as each other. The total **if** construct is itself an expression therefore of type Real.

For those unfamiliar with the languages in which this sort of construction is allowed, since it is an expression, it can for example be placed on the right-hand side of an assignment statement

$x := $ **if** B **or** C **then** $x + 7.0$ **else** $x - 7.0$

or as a parameter in a procedure or function call

SIN (**if** B **or** C **then** $x + 7.0$ **else** $x - 7.0$)

The '**if**' construct illustrated thus takes two Booleans and a Real and produces a Real result. Its type is therefore

$$\mathbb{B} \times \mathbb{B} \times \mathbb{R} \to \mathbb{R}$$

Example 6.16: Telephone Accounts System

This example, introduced in chapter 2 and continued in chapter 4, can be made more exact with the help of mappings.

In chapter 4 we said that a subscriber could be represented by a tuple which was a member of the Cartesian product

$$C^* \times B^* \times TN$$

where C^* was the set of character strings representing names, B^* the set of character strings representing addresses, and $TN = D^4$ was the set of quartuples of digits representing four-digit telephone numbers. A problem with this representation is that two different subscribers—that is, two different members of $C^* \times B^* \times TN$—could have the same telephone number. Thus for example

$\langle$'J. Smith', '2 Ash Grove', 3768$\rangle$

$\langle$'F. Brown', '37 Lime St.', 3768$\rangle$

could represent two separate subscribers in that scheme. A mapping can resolve this:

$$TN \to C^* \times B^*$$

6.7 DEFINING FUNCTIONS

In the contexts of computing and programming, the values of data types which we wish to manipulate need in general to be finite objects. Hence we distinguish between 'functions', which may have finite or infinite domains, and 'mappings', which are functions with finite domains only. A mapping can thus be represented by a table (implemented as an array perhaps). We typically implement functions by means of algorithms which, given any values of the function's arguments, produce as result the value of the corresponding images.

Tables

Definition 6.15
A *Table* is a finite graph, and hence can define a finite function by means of its graph.

Example 6.17

The Boolean operation $\wedge$ is a function

$$\wedge : \mathbb{B} \times \mathbb{B} \to \mathbb{B}$$

and can be defined for each value of its domain

True $\wedge$ True $=$ True
True $\wedge$ False $=$ False
False $\wedge$ True $=$ False
False $\wedge$ False $=$ False

Thus can be conveniently represented by means of the familiar table

x	y	$x \wedge y$
T	T	T
T	F	F
F	T	F
F	F	F

Similarly multiplication over the numbers modulo 3

x	y	$x * y$
0	0	0
0	1	0
0	2	0
1	0	0
1	1	1
1	2	2
2	0	0
2	1	2
2	2	1

Algorithms

However, functions of infinite domains cannot be defined in this way. Multiplication of natural numbers used to be taught by means of tables, but only a finite subset of the natural numbers (0 to 10 or 0 to 12) could be covered in this way. To define multiplication over the rest of the natural numbers, another technique is needed, and the traditional means used in teaching multiplication to children is to use an algorithm. The algorithm consists of a scheme of mechanical instructions whereby any two numbers of any finite number of digits can be processed to give a result which is their product. The scheme consists of iteratively applying the table to pairs of digits, adding in a carry digit, etc.

Definition 6.16

An *algorithm* is a procedure which terminates, and a *procedure* is a sequence of mechanically executable steps. In particular, a procedure in a high level language (which terminates) is an example of an algorithm.

   Algorithms are rather an unsatisfactory way of defining a function. Firstly, for any given function there is a wide variety of algorithms. Secondly, while an algorithm is a statement of how a computation is carried out, it does not state what the computation does, and it can be by no means clear what is the effect of the computation.

Example 6.18

To illustrate the first point, let us consider a number of algorithms for calculating a factorial.

(i)   fact: $\mathbb{N} \to \mathbb{N}$

   fact (x): **if** $x = 0$ **then** fact$:= 1$

                        else fact$:=$ fact $(x - 1) * x$

The above is called a 'recursive' algorithm because it invokes itself. It is perhaps the clearest of the three algorithms illustrated because its structure reflects closely a recursive definition of the function 'fact' which could be defined

   fact $(x) =$ **if** $x = 0$ **then** 1

                        else fact $(x - 1) * x$

(ii)   fact (x): **Int** temp, result;

                   temp$:= 0$; result$:= 1$;

                   While temp $< x$

                   do temp$:=$ temp $+ 1$;

                        result$:=$ result $*$ temp;

                   od;

         fact$:=$ result;

This algorithm uses iteration rather than recursion, and its structure is, of course, completely different, using local variables and a loop. How could we show that this algorithm produces the same result for all values of x as the first? We could test the two, but never exhaustively since there are an infinite number of values of x for which the algorithms could be tested. Some knowledge of the interior of the algorithms is necessary in order to be convinced that they work correctly. We shall see this in more detail in chapter 11.

(iii)   fact (x): **array** $[0:256]$

Int A **init** $(1, 1, 2, 6, 24, 120, \ldots)$;

if $x > 256$ **then** cause overflow

else fact$:= A[x]$;

This last algorithm does actually use a finite table, relying on the fact that factorials get rapidly very large and no computer could represent with normal integer representation a number of the size of 256!.

These examples illustrate how a wide variety of algorithms can define the same function.

The second point, that an algorithm does not necessarily show clearly what function it implements, is best illustrated by an exercise. The following is attributed to E.W. Dijkstra.

Exercise 6.3

The following is a description of an algorithm. One starts with an opaque bag containing b black balls and w white balls (these constitute the input data). One has an unlimited stock of white balls (a local variable). The algorithm consists of a sequence of moves, until there is only one ball left. The moves are as follows (see figure 6.18)

Extract two balls from the bag

If they are both the same colour, put back one white ball (taking one from the stock if two black walls were extracted)

If they are different colours, put back one black ball

Since each move extracts two balls from the bag and replaces one, the number of moves is bounded and the process will terminate when just one ball is left in the bag.

The algorithm thus starts with an input of two natural numbers b and w, and produces as result a binary value of type {black, white}. The question is how is the result related to the input, or what function of (b, w) does the algorithm implement? *Hint:* there is clearly a non-deterministic element to the process in that, if there is a mixture of colours in the bag,

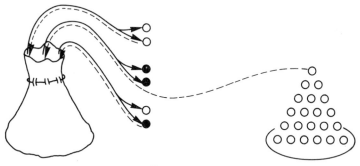

Figure 6.18

extracting two balls may produce a variety of results. Before assuming that the whole algorithm produces a non-deterministic result and considering statistical ideas, you are advised to try out the process on paper with 2, 3 and possibly 4 balls in the bag.

The point which the above exercise illustrates is that a conceptually opaque algorithm may implement a conceptually simple function, and therefore algorithms are not the most satisfactory way of defining functions.

Explicit Definitions

We have seen one sort of explicit definition of a function, namely by means of explicitly defining its value for each value of its arguments, often conveniently arranged as a table. This is only possible for finite functions however. We can also give an explicit definition of a function in terms of other functions which are already defined.

Examples 6.19

(i) Square: $\mathbb{R} \to \mathbb{R}$

 Square $(x) \triangleq x * x$

(ii) Fact: $\mathbb{N} \to \mathbb{N}$

 fact $(x) \triangleq$ **if** $x = 1$ **then** 1

 else $x * $ fact $(x - 1)$

These two definitions assume that $*$ is defined already. The second uses recursion, that is in its definition it makes use of the function 'fact' again. In any recursive definition there must always be a 'path' which does not call the function being defined. In the above example this path is followed in the case that $x = 1$. Further, all other cases must eventually lead to invoking the function so that a non-recursive path is taken. Note in these definitions the use of the $\triangleq$ symbol to mean 'is equal to by definition'.

At the end of chapter 5, some useful operations of functions on lists were defined (an operation is actually the same thing as a function). We are now in a position to define some of these more rigorously.

(iii)   dels: $A* \times \mathbb{N} \to A*$

dels $(l, n) \triangleq$ **if** $n >$ len l **then** l

        else if $n = 1$ **then** tl l

                else $\langle$hd $l\rangle \| (\text{tl } l, n - 1)$

(iv)   mods: $A* \times \mathbb{N} \times A \to A*$

mods $(l, n, a) \triangleq$ **if** $n >$ len l **then** l

      else if $n = 1$ **then** $\langle a \rangle \|$ tl l

           else $\langle$hd $l\rangle \|$ mods (ti $l, n - 1, a$)

(v)   front: $A* \times \mathbb{N}_0 \to A*$

front $(l, k) \triangleq$ **if** $k >$ len l **then** l

      else if $k = 0$ **then** $\langle\ \rangle$

           else $\langle$hd $l\rangle \|$ front (tl $l, k - 1$)

Exercise 6.4

Define by explicit definition the subs function as given at the end of chapter 5. *Hint:* you may find it useful to make use of the front function above.

Pre-conditions

In the above example, default values are supplied as results of the functions in cases of anomalous input values. For example, in dels (l, n) if n is greater than the length of the string, the function delivers the original string unchanged. It can sometimes be artificial to supply default values in such cases, and an alternative technique is to define a partial function whose domain does not include these unwanted values of the domain.

This can be done by defining a suitable type, but doing so can lead to notational clumsiness. For in the dels example we would have to define the input as a single argument being members of a Cartesian product which satisfy the appropriate condition

$$T = \{\langle l, n \rangle | \langle l, n \rangle \in (A* \times \mathbb{N}) \wedge n < \text{len } l\}$$
$$\text{dels}: T \to A*$$

The type T is now the set of all pairs of A-sequences and natural numbers such that the Natural number does not exceed the length of the sequence.

An alternative is to specify a pre-condition. This states that the function is partial and is defined only for those values of the input argument which satisfy the pre-condition.

Definition 6.17
A *pre-condition* for a function is a predicate applied to the arguments of
the function which restricts the domain to those values of the arguments
which give a value of True for the predicate.

Example 6.20

Using this approach dels would be defined as follows

$$\text{dels}: A^* \times \mathbb{N} \to A^*$$

$$\text{pre-dels}: A^* \times \mathbb{N} \to \mathbb{B}$$

$$\text{pre-dels}\,(l, n) \triangleq n < \text{len } l$$

$$\text{dels}(l, n) \triangleq \textbf{if } n = 1 \textbf{ then } \text{tl } l$$
$$\textbf{else } \langle \text{hd } l \rangle \sqsubseteq \text{dels}(\text{tl } l, n-1)$$

Observe that a type clause is given for the pre-condition. The pre-
condition is itself a function, taking the same arguments as the function
being defined and delivering True or False. Note the form of the function
definition: all the symbols representing variable values are bound to
occurrences of those symbols in the input parameters. Thus

n and l are bound to their occurrences as input parameters on the left of
the $\triangleq$ symbol, and have no connection with any occurrences of n and l
outside the definition. They have meaning only within the scope of the
definition. Hence the above definition is exactly the same as, and can be
used in exactly the same contexts as, the following

$$\text{dels}(p, x) \triangleq \textbf{if } x = 1 \textbf{ then } \text{tl } p$$
$$\textbf{else } \langle \text{hd } p \rangle \| \text{dels}(\text{tl } p, x - 1)$$

Thus it was not necessary (although it perhaps makes for greater clarity)
to use the same symbol l and n in the definition of dels and pre-dels.

Exercise 6.5

Restate the definition of mods, front and subs using pre-conditions to
limit the domains instead of producing default results, in cases where the
numeric parameter exceeds the length of the string.

Notation

Since the type of clause for the pre-condition is invariably the same as the type of clause of the function with the type of the result replaced by $\mathbb{B}$, it is frequently omitted.

Another way of defining functions, by means of their *implicit specifications*, is explained in chapter 10.

Exercise 6.6 (Workshop)

In the space-borne computer problem (see exercise 3.10) recall the data

position	(x, y, z)
velocity	$(\dot{x}, \dot{y}, \dot{z})$
acceleration	$(\ddot{x}, \ddot{y}, \ddot{z})$
attitude	$(p, q, r), 0 \leqslant p, q, r \leqslant 2\pi$
mass	m
motors	M (finite set)
burn rate	$\dot{e}_f$
duration	d_f

also the position and masses of heavenly bodies

$$s_E, m_E$$

We used subscripts 'E', 'M', 'f' to represent what we can now recognise are mappings. Thus if H is a finite set of heavenly bodies, we have mappings

$$S:H \rightarrow pos$$
$$m:H \rightarrow mass$$

where $pos = \mathbb{R}^3$, $mass = \{x \in \mathbb{R} | x \geqslant 0\}$, for example.

Re-examine the data of this problem to see how mappings can be used to express uniqueness of association of quantities.

Exercise 6.7 (Workshop)

In the fault-tolerant control system in exercise 3.11, subscripts were again used to associate a function f_b and a checker c_b with each of a set of functional blocks F. Use mapping notation to describe the system without the use of subscripts.

Exercise 6.8 (Workshop)

In the personal diary problem (see exercises 4.5 and 3.4), many mappings are inherent in the problem data. There is a mapping from each date to a set of events; from each event to a start time, a duration and a description. Restate the problem using mappings, and write a set-expression for the set of events occurring on a given day, and at a given time on a given day. Formulate an invariant condition, which will be a logical expression, which expresses the constraint that no two events in the diary may overlap.

Exercise 6.9 (Workshop)

In the nuclear power station exercise (see exercises 4.4 and 3.8) there is also much scope for the use of mappings. Each sensor S is associated with a sensor type belonging to $\{T, E, CT, CP, CF\}$. Express the sensors and their outputs using mapping notation. Likewise restate the positions of the rods, the association between sensors and dials and lamps, and the association between the set V of VDUs and the current message being displayed on them using mapping notation. The details of types etc. are given in exercise 3.8.

   This is a lengthy but rewarding exercise especially suitable for supervised group work. Several iterations of the solution may be anticipated.

Exercise 6.10 (Workshop)

If you have been using an 'own problem' as suggested in chapter 1, it is very instructive at this stage to consider how it may be formulated in terms of mappings.

SUMMARY OF RESULTS

Result	Reference
The set $\mathbb{Z}$ of integers is countable	Theorem 6.1
The even natural numbers are countable	Theorem 6.2
The set of all pairs of natural numbers is countable (hence the rationals $\mathbb{Q}$ are countable)	Theorem 6.3
The reals $\mathbb{R}$ are uncountable	Theorem 6.4

(continued overleaf)

Summary of Results (continued)

Result	Reference
The set of all mappings $\mathbb{N} \to \mathbb{N}$ is uncountable	Theorem 6.5
$\mathscr{P}(\mathbb{N})$ is uncountable	Theorem 6.6
$(M^{-1})^{-1} = M$	Theorem 6.7

Chapter 7 Relations

7.1 DEFINITIONS

There are several equivalent ways of defining relations. The term itself is suggestive in that it evokes the notion of a relationship between two entities. Thus in a population we may have various 'relations' existing: cousins, uncles, parents, ancestors etc.

One way of defining a relation is as a predicate function of two variables.

Definition 7.1
A *binary relation* R is a predicate of two arguments

$$R: A \times B \rightarrow \mathbb{B}$$

147

Example 7.1

Continuing our population analogy

   Is-parent:$Pop \times Pop \rightarrow \mathbb{B}$
   Is-descended-from:$Pop \times Pop \rightarrow \mathbb{B}$
   Is-sibling:$Pop \times Pop \rightarrow \mathbb{B}$

are all relations. So for example

   Is-parent (a, b)

takes the value True when a is the parent of b.

   A second way of defining a relation is as a generalisation of a mapping. A mapping M has a graph which is a set of pairs $(a, b) \in A \times B$ such that

   $$\forall a, a' \in A, b \in B . ((a, b) \in M \land (a', b) \in M \Rightarrow a = a')$$

If we lift the constraint we have a relation; thus a relation can be defined simply as a set of pairs.

Definition 7.2
A *binary relation* R' is a subset of a Cartesian product of two sets A, B

   $$R' \subseteq A \times B$$

Suppose R is a relation expressed as

   $$R : A \times B \rightarrow \mathbb{B}$$

and R' is the same relation expressed as a set of pairs

   $$R' \subseteq A \times B$$

   then

   $$R(a, b) \Leftrightarrow (a, b) \in R'$$

Notation

The usual notation for a relation is to write

   aRb

meaning 'a is in relation R to b' or $R(a, b)$ or $(a, b) \in R$. Hence we would normally write

   a *Is-parent* b

etc.

Definition 7.3

If a relation R is defined on two sets which are the same, that is $R \subseteq A \times A$ or $R : A \times A \to \mathbb{B}$, the relation is said to be *on* A. Thus the relations *Is-parent* etc. are 'on' *Pop*.

Graphs of Relations

With mappings or functions we can define graphs. We can similarly define graphs of relations since relations are generalisations of mappings.

Definition 7.4

The *graph of a relation* R is the set of pairs of points (a, b) in the co-ordinate space $A \times B$ for which aRb. As indicated by definition 7.2, a relation is often simply identified with its graph.

Example 7.2

(i) Suppose our relation is on $\mathbb{R}$ and is the relationship of equality. Then the graph of $=$ is as shown in figure 7.1. The graph of the relation $>$ is shown in figure 7.2.

(ii) Consider the powerset P of $\{a, b, c\}$ namely

$$\{A, B, C, D, E, F, G, H\}$$

where

$$A = \{\ \}$$
$$B = \{a\}$$
$$C = \{b\}$$
$$D = \{c\}$$
$$E = \{b, c\}$$
$$F = \{a, c\}$$
$$G = \{a, b\}$$
$$H = \{a, b, c\}$$

Then the relation $\subseteq : P \times P \to \mathbb{B}$ has a graph as follows

	A	B	C	D	E	F	G	H
A	X							
B	X	X						
C	X		X					
D	X			X				
E	X		X	X	X			
F	X	X		X		X		
G	X	X	X				X	
H	X	X	X	X	X	X	X	X

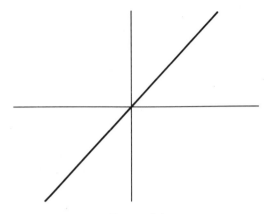

Figure 7.1

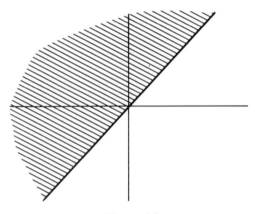

Figure 7.2

Exercise 7.1

Sketch the graphs of the following relations on R

(i) $xRy \Leftrightarrow x+y \geqslant 1$
(ii) $xRy \Leftrightarrow x+y \leqslant 4$
(iii) $xRy \Leftrightarrow x+y \geqslant 4$

Exercise 7.2

What is the relation whose graph is shown in figure 7.3?

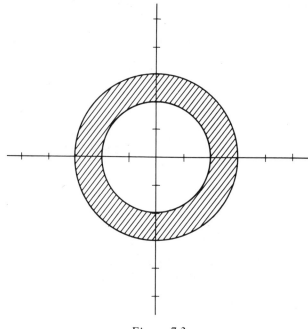

Figure 7.3

Example 7.3: Telecommunications

In our telephone accounts problem, each subscriber may have a number of features supplied

> *Features* = {Ordinary extensions,
> Socket extensions,
> Push-button dials,
> Abbreviated dialling,
> Automatic alarm call,
> Message recording}

A subscriber may have none, several, or even all of these features. This could be represented in the database by a relation

> *Has-features* ⊆ *Subs-id* × *Features*

Then if

> *s Has-features f*

it means that subscriber *s* has the feature *f*.

Example 7.4: Personnel File

In our Personnel file example, there may be a number of attributes

> $Perks = \{$Company car,
> Senior employees' pension,
> Loans available,
> Medical insurance,
> Expense account$\}$

Again the possession of any or several or all these facilities may be represented in the database by a relation

> $Has\text{-}unfair\text{-}privilege \subseteq Employee\text{-}id \times Perks$

so that

> E126 *Has-unfair-privilege* Company car

means that the employee with identity E126 has a company car.

That employee may also have other perks; for example, the expression

> E126 *Has-unfair-privilege* Medical insurance

could also have the value True.

The fact that, in both these examples, a subscriber may have zero or several features, and an employee may have zero or several perks, means that these attributes cannot be expressed as functions, but must be relations. However, whenever we have a relation $R \subseteq A \times B$ we can in fact express this as a function $f_R : A \to \mathscr{P}B$ which would be defined

$$f_R(a) \triangleq \{x \mid x \in B \wedge aRx\}$$

Thus in the example with $\subseteq : \mathscr{P}\{a,b,c\} \times \mathscr{P}\{a,b,c\} \to \mathbb{B}$ we would have

$$f_c(e) \triangleq \{x \mid x \in \mathscr{P}\{a,b,c\} \wedge e \subseteq x\}$$

so

$$f_c(\{\ \}) = \{\{\ \}, \{a\}, \{b\}, \{c\}, \{b,c\}, \{a,c\}, \{a,b\}, \{a,b,c\}\}$$
$$f_c(\{a\}) = \{\{a\}, \{a,b\}, \{a,c\}, \{a,b,c\}\}$$
$$f_c(\{b,c\}) = \{\{b,c\}, \{a,b,c\}\}$$
$$f_c(\{a,b,c\}) = \{\{a,b,c\}\}$$

etc.

Exercise 7.3

Take the relation found in exercise 7.2 and re-express as a function of type $\mathbb{R} \to \mathscr{P}\mathbb{R}$.

7.2 PROPERTIES OF RELATIONS

For relations on a set A—that is, for relations $R \subseteq A \times A$—there are a number of useful and interesting properties.

Definition 7.5
A relation R is *reflexive* if, $\forall a \in A \,.\, aRa$. Thus if every member of the set A is in relation R to itself, R is reflexive.

Definition 7.6
A relation R is *symmetric* if, whenever we have aRb then we also have bRa. Formally, R is symmetric if

$$\forall a, b \in A \,.\, aRb \Rightarrow bRa$$

Definition 7.7
A relation R is *transitive* if, whenever we have aRb and bRc then we also have aRc. Formally, R is transitive if

$$\forall a, b, c \in A \,.\, aRb \wedge bRc \Rightarrow aRc$$

Example 7.5

$\subseteq$ is reflexive because for all sets s, $s \subseteq s$.
$\subseteq$ is transitive because for all sets s_1, s_2, s_2, if $s_1 \subseteq s_2 \subseteq s_3$ then $s_1 \subseteq s_3$.
$\subseteq$ is not symmetric however.

$\leqslant$, like $\subseteq$, is reflexive, transitive but not symmetric. $=$ is reflexive, transitive and symmetric.

Exercise 7.4

State whether the following relations are reflexive, transitive, symmetric

(i)   On a population, the relation 'is a sibling of'.
(ii)   On a population, the relation 'is a sister of'.
(iii)   On a population, the relation 'is an ancestor of'.
(iv)   'is a parent of'.
(v)   'is a descendant of'.
(vi)   'is a first cousin of'.
(vii)   'is a blood relation of'.
(viii) On $\mathbb{Z}$, $<$.
(ix)   On $\mathbb{Z}$, $\neq$.

7.3 EQUIVALENCE RELATIONS

Definition 7.8
An *equivalence relation* is a relation which is reflexive, transitive and symmetric. It captures some notion of equivalence, sameness, or similarity.

Example 7.6

Thus, on $\mathbb{N}$, the following is a equivalence relation

'Has the same prime factors'

For $N \geqslant 2$, congruence modulo N on the integers is also an equivalence relation; thus

$$a \approx_N b \triangleq \mathrm{REM}(a, N) = \mathrm{REM}(b, N)$$

where the function REM (remainder) is defined such that

$$0 \leqslant \mathrm{REM}(x, n) < n$$

and

$$x = k * n + \mathrm{REM}(x, n)$$

for some integer k. (We shall provide a notation for defining such functions more precisely in chapter 10.)

Thus if we take $N = 3$ the following groups of numbers are all congruent to each other modulo N

$$\{0, 3, 6, 9, \ldots, -3, -6, -9, \ldots\}$$
$$\{1, 4, 7, 10, \ldots, -2, -5, -8, \ldots\}$$
$$\{2, 5, 8, 11, \ldots, -1, -4, -7, \ldots\}$$

If we call these groups of numbers [0], [1], [2], respectively, then for any member x of [0], $\mathrm{REM}(x, 3) = 0$, for any member x of [1], $\mathrm{REM}(x, 3) = 1$ and for any member x of [2], $\mathrm{REM}(x, 3) = 2$.

Thus the equivalence relation $\approx_3$ partitions the integers into three disjoint subsets. The result is general.

Definition 7.9
Given an equivalence relation E over a set A, if $a \in A$ then the set of all elements x of A such that xEa (and hence because of symmetry, such that aEx) is written

$$[a]$$

and is called the *equivalence class* of a.

Theorem 7.1
An equivalence relation E on a set A partitions A into disjoint subsets
called *equivalence classes*.

Proof
Suppose $[a]$ and $[b]$ are two such subsets of A.
Either $[a] \cap [b] = \varnothing$ or $\exists x \in A . x \in [a] \land x \in [b]$
If $\exists x \in A . x \in [a] \land x \in [b]$
then $\forall y \in [a] . yEx$
and $\forall z \in [b] . xEz$
that is $\forall y \in [a], \forall z \in [b] . yEx \land xEz$
but because E is transitive we can deduce

$$\forall y \in [a], \forall z \in [b] . yEz$$

hence $[a] = [b]$
Therefore either $[a] = [b]$ or $[a] \cap [b] = \varnothing$

Theorem 7.2
Conversely, if we have a set A partitioned into a collection of disjoint
subsets, we can derive from the partition an equivalence relation.

Proof
Let P be the partition; that is

$$P \subseteq \mathscr{P}A \land (\forall a \in A . \exists p \in P . a \in p)$$
$$\land (\forall s, t \in P . s = t \lor s \cap t = \varnothing)$$

then we can define an equivalence relation

$$E_P \subseteq A \times A$$

such that

$$aE_Pb \Leftrightarrow (\exists p \in P . a \in p \land b \in p)$$

Then P is precisely the set of equivalence classes of the equivalence
relation E_P.
   We can show that E is indeed an equivalence relation—that is,
reflexive, transitive and symmetric—simply as follows.
Reflexivity. From the definition of partition used above

$$\forall a \in A . \exists p \in P . a \in p$$

hence

$$\forall a \in A . \exists p \in P . (a \in p \land a \in p)$$

hence

$$\forall a \in A . aE_Pa$$

from the definition of E_P.

Symmetry. This follows from the definition of E again and the symmetry of $\wedge$.

Transitivity. If $a \in A$ and $a \in p1 \in P$ and $a \in p2 \in P$ then we can deduce from the second clause in the definition of partition above that $p1 = p2$. Hence if aE_pb and bE_pc we have

$$(\exists p \in P . (a \in p \wedge b \in p)) \wedge (\exists q \in P . (b \in q \wedge c \in q))$$

but from the above we know that $p = q$, so that

$$\exists p \in P . (a \in p \wedge b \in p \wedge c \in p)$$

hence aE_pc.

SUMMARY OF RESULTS

Result	Reference
An equivalence relation generates equivalence classes	Theorem 7.1
Converse of theorem 7.1	Theorem 7.2

PART III: Algebraic Concepts

In Parts I and II we have seen how to use the language of set theory and extensions to it to define data in abstract terms. Although data is the 'subject matter' of a program, a useful program has more to it, namely procedures, functions, etc., expressed as algorithms if a conventional algorithmic language is used.

Programs are contemporarily described as 'Data + Algorithms'. An algebra could be considered as 'Sorts (or Types) + Operations'. The sorts in an algebra are the types composed of the set-theoretic constructs we have already met, and are an abstraction of the program data. The program data is a model of the types or sorts. Likewise the operations in the algebra are an abstraction of the algorithmic processes (using this term in a broad sense) which the program can carry out. The 'algorithms' of the program are thus a model of the algebraic operations.

The ideas of algebra in chapter 8 enable us to describe more completely what a program does, in abstract terms, that is without prejudice as to how it does it. Algebra therefore give us an ability to describe the intended actions of programs yet to be written, and hence provides the basis for specifications of programs. Formulating the specification of a program is a stage in the software engineering development process which should be undertaken prior to design and coding, as is well accepted. Specifications are dealt with in more detail in Part IV.

Chapter 9 deals with homomorphisms. Homomorphisms are particular kinds of relationships between algebras; they provide us with the criteria we need for determining whether a program is a correct model of the algebra which is its specification. Certain kinds of homomorphisms, called 'forgetful', can be related to some software engineering ideas such as data security, access control, classes of users of a system etc.

Chapter 10 concerns functions in an algebra, which are in effect subsidiary or derived operations, and can represent abstractions of programming procedures, functions, and other units.

Thus Part III completes the ideas we need for formulating abstractions of programs, which are essential for the processes of producing specifications and showing that programs meet them.

Chapter 8 Algebras

NOTATION INTRODUCED

Notation	Concept	Reference
	dyadic operation	Def. 8.1
	adicity, arity	Def. 8.2
	identity element	Def. 8.3
	overloading	Section 8.2
	associativity	Def. 8.4
$a \,\square\, b \,\square\, c$	dropping brackets	Section 8.3
$-a, a^{-1}$	inverses	Def. 8.5
	semigroup, monoid, group	Def. 8.6
	commutative, Abelian	Def. 8.7
	closure of an operation	Def. 8.8
	rings	Def. 8.9
	fields	Def. 8.10
	homogeneous algebra	Def. 8.11
	signature, sort, scheme, heterogeneous, many-sorted	Def. 8.12

This chapter develops the mathematics of simple algebras. The initial part of the development may seem at first to have no connection to the problems of programming and software engineering. I ask readers to bear with me: the developments later in the chapter, in which our now familiar running examples, and finally our workshop exercises, are related to applications. The application of the theory is delayed only because the application requires a certain depth of development. The initial work has therefore to be done in a somewhat abstract context.

8.1 OPERATIONS

Definition 8.1
A total mapping $M: A \times A \to A$ is called a *dyadic operation*.

Example 8.1

Here are some familiar examples.

(i) $+: \mathbb{Z} \times \mathbb{Z} \to \mathbb{Z}$

$-: \mathbb{Z} \times \mathbb{Z} \to \mathbb{Z}$

$*: \mathbb{Z} \times \mathbb{Z} \to \mathbb{Z}$

$/: \mathbb{Z} \times \mathbb{Z} \to \mathbb{Z}$

(ii) $+: \mathbb{R} \times \mathbb{R} \to \mathbb{R}$

etc.

(iii) $\wedge: \mathbb{B} \times \mathbb{B} \to \mathbb{B}$

$\vee: \mathbb{B} \times \mathbb{B} \to \mathbb{B}$

$\Rightarrow: \mathbb{B} \times \mathbb{B} \to \mathbb{B}$

The term 'dyadic' indicates that the operation takes two arguments. An operation may not be dyadic. Examples of monadic operations are

$-: \mathbb{Z} \to \mathbb{Z}$

$\sim: \mathbb{B} \to \mathbb{B}$

$-: \mathbb{R} \to \mathbb{R}$

Definition 8.2
The *adicity* or *arity* of an operation is the number of arguments it takes. Sometimes constants and variables may be conveniently regarded as operations of zero adicity.

Examples 8.2

Thus in arithmetic or Boolean algebra we can regard an operation like '+' as a mapping which, given any pair of elements, produces a new value: see figure 8.1.

In general, operators may take arguments of different types. The if...then...else... construct as used in ALGOL 60 and other languages takes a Boolean argument and two arguments of another type, and produces a value which is of the same type as the second and third argument. Thus

'if-then-else': $\mathbb{B} \times X \times X \to X$

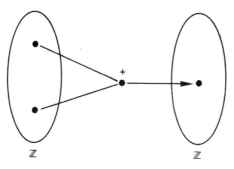

Figure 8.1

where X is some type. 'if-then-else' has adicity 3.

In general, an operation of adicity n maps an n-tuple to a value of a further type, thus

$$f : A_1 \times A_2 \times \ldots \times A_n \to Y$$

or

$$f : \overset{n}{\underset{1}{\times}} A_i \to Y$$

The operations with which we are most familiar are usually expressed in infix notation, where the 'name' of the operator is placed between its arguments

$$a + b$$
$$S \wedge T$$

However, this is difficult to arrange if there are more than two arguments

$$f(a, b, c)$$

and in fact the more familiar operations can be used in this general way in what is known as Polish notation

$$+ ab$$
$$- cd$$

so that

$$* + ab + cd$$

means

$$(a + b) * (c + d)$$

in infix notation. Polish notation permits one to avoid the use of brackets.

Thus, in general, any operation is a mapping from the Cartesian

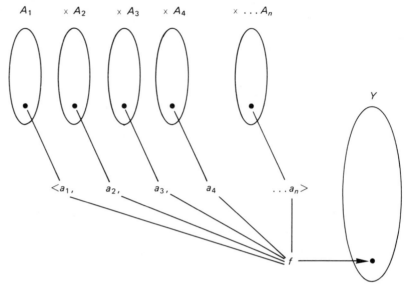

Figure 8.2

product of the types of its arguments to the type of its result. Its arguments may be considered as a tuple: see figure 8.2.

8.2 IDENTITIES

Definition 8.3
With some dyadic operations there is associated an identity element. If F is an infix dyadic operation $F:A \times A \to A$ then $i \in A$ is an *identity element* for F if

$$\forall a \in A . aFi = a = iFa$$

Theorem 8.1

For any dyadic operation there is at most one identity element.

Proof
Suppose that $i \in A$ and $j \in A$ are both identities for F; then

$$i = jFi = j$$

□

Example 8.3

The identity for $+:\mathbb{Z} \times \mathbb{Z} \to \mathbb{Z}$ is 0.
The identity for $*:\mathbb{Z} \times \mathbb{Z} \to \mathbb{Z}$ is 1.
The identity for $\wedge :\mathbb{B} \times \mathbb{B} \to \mathbb{B}$ is T.
The identity for $\vee :\mathbb{B} \times \mathbb{B} \to \mathbb{B}$ is F.

Exercise 8.1

Given a 'universal set' X, the usual union operation is defined

$$\cup:\mathscr{P}X \times \mathscr{P}X \to \mathscr{P}X$$

What is the identity for $\cup$?

Exercise 8.2

If the intersection operator $\cap$ is similarly defined

$$\cap:\mathscr{P}X \times \mathscr{P}X \to \mathscr{P}X$$

what is its identity?

Exercise 8.3

Does

$$-:\mathbb{Z} \times \mathbb{Z} \to \mathbb{Z}$$

have an identity?

Exercise 8.4

Does

$$\Rightarrow:\mathbb{B} \times \mathbb{B} \to \mathbb{B}$$

have an identity?

Notation

Certain operator symbols, such as $+$ and $*$, are conventionally
'overloaded'; that is, we accept the idea that we may have two or more
quite separate operations

$$+:\mathbb{Z} \times \mathbb{Z} \to \mathbb{Z}$$
$$+:\mathbb{R} \times \mathbb{R} \to \mathbb{R}$$
$$+:\mathbb{N} \times \mathbb{N} \to \mathbb{N}$$

Similarly for $*$. By convention we denote the identity for any operation named '$+$' by '0', and we denote the identity for any operation named '$*$' by '1'.

More Examples 8.4

(i) Strings
Given an alphabet A, consider the set of all strings over that alphabet $A*$. There is the usual concatenation operator:

$$\| : A* \times A* \to A*$$

This has an identity element which consists of the empty string $\langle \ \rangle$.
(ii) Transactions on a file
A particular kind of alphabet and strings over it are the transactions which may be performed on a file. The 'particulate' transactions which cannot be divided up form the alphabet, and a general transaction is an ordered sequence of particulate transactions. The identity will be the null transaction I, which does not alter the file. This, when concatenated with any other transaction T, either $I\|T$ or $T\|I$, produces the same transaction T again.

Of course certain sequences of transactions may be the same as each other; that is, if A, B, C, D are particulate transactions, it may be that

$$A\|B = C\|D$$

just as, back in the realm of numbers

$$1 + 3 = 2 + 2$$

Such equations form the 'theorems' of the algebra. A 'theorem' in any algebra is actually any 'true' statement about terms (values or variables) in the algebra. Thus in the algebra of numbers, '$2 + 2 = 4$' is a theorem, although it is so basic one may not usually think of it as such. A more substantial theorem in the algebra of numbers might be

If $x + a = y + a$

then $x = y$

and in the algebra of strings (or transactions on a file)

If $x\|a = y\|a$

then $x = y$

8.3 ASSOCIATIVITY

Definition 8.4
An operation $\square : A \times A \to A$ is *associative* if

$$\forall a, b, c \in A \, . \, a \square (b \square c) = (a \square b) \square c$$

Notation

When an operation is associative, we can drop the brackets when writing a sequence of elements separated by the operator symbol

$$a \,\square\, b \,\square\, c \,\square\, d$$

because there is no ambiguity: we can place any pairs of matching brackets in the sequence of symbols and obtain the same value

$$((a \,\square\, b) \,\square\, c) \,\square\, d$$
$$= (a \,\square\, (b \,\square\, c)) \,\square\, d$$
$$= a \,\square\, ((b \,\square\, c) \,\square\, d)$$
$$= a \,\square\, (b \,\square\, (c \,\square\, d))$$
$$= (a \,\square\, b) \,\square\, (c \,\square\, d)$$

and so on for a sequence of any length

Exercise 8.5

Which of the following operations are associative?

(i) $+$
(ii) $-$
(iii) $*$
(iv) $/$
(v) $\cup$ with $\mathscr{P}(S)$
(vi) $-$ with $\mathscr{P}(S)$
(vii) $\cap$ with $\mathscr{P}(S)$
(viii) $\vee$ with $\mathbb{B}$
(ix) $\wedge$ with $\mathbb{B}$
(x) $\Rightarrow$ with $\mathbb{B}$

Theorem 8.2

Strings over an alphabet A, with the concatenation operator $\|$ are associative.

Proof
If we have strings S_1, S_2, S_3 of lengths l_1, l_2, l_3 respectively, the length of

$$S_1 \| (S_2 \| S_3)$$

is

$$l_1 + (l_2 + l_3) = l_1 + l_2 + l_3$$

and the length of

$$(S_1 \| S_2) \| S_3$$

is

$$(l_1 + l_2) + l_3 = l_1 + l_2 + l_3$$

and in each the ith element is

—the ith element of S_1 if $i \leqslant l_1$
—the $(i - l_1)$th element of S_2 if $l_1 < i \leqslant l_1 + l_2$
—the $(i - l_1 - l_2)$th element of S_3 if $l_1 + l_2 < i \leqslant l_1 + l_2 + l_3$ $\square$

8.4 INVERSES

Definition 8.5: Inverses
An element a of a set A has an *inverse* b under an operation $\square$ if

$$a \square b = b \square a = i$$

where i is the identity element.

Notation

For any operation called '$+$' the inverse of 'a' is written '$-a$'.
For any operation called '$*$' the inverse of 'a' is written 'a^{-1}'.

Example 8.5

(i)   Every element of the integers with $+$ has an inverse.
(ii) Every element of the reals except 0 with $*$ has an inverse.
(iii) Strings over an alphabet with $\|$ do not have inverses: given a string
      there is no string S_2 such that

$$S_1 \| S_2 = \langle \ \rangle$$

      unless S_1 is itself the null string.
(iv) Subsets of a set S do not in general have inverses with respect to
      either $\cup$ or $\cap$.

Exercise 8.6

(i)   Recalling that T is an identity for $\wedge$, do the Boolean truth values
      have inverses with respect to $\wedge$?
(ii) Recalling that F is an identity for $\vee$, do the Boolean truth values
      have inverses with respect to $\vee$?

8.5 GROUPS

Definitions 8.6
A set with a total associative operation is called a *semigroup*.
A semigroup with a identity element is called a *monoid*.
A monoid in which every element has an inverse is called a *group*.

Exercise 8.7

Are the following semigroups, monoids, groups or none of these?

(i)   The integers with $+$.
(ii)   The integers with $-$.
(iii) The integers with $*$.
(iv)   The integers with $/$.
(v)   The reals with $+, -, *, /$ respectively.
(vi)   The booleans with $\vee, \wedge, \Rightarrow$ respectively.
(vii) $\mathscr{P}(S)$ where S is a given non-empty set, with $\cup, \cap$ respectively.

Theorem 8.3

In any group with an operation $+$

$$x + a = b$$
$$a + y = b$$

have unique solutions for x and y respectively

$$x = b + (-a) \text{ and } y = (-a) + b$$

Hence

$$c + a = d + a \Rightarrow c = d$$

likewise

$$a + c = a + d \Rightarrow c = d$$

(This is called the cancellation law.)

Proof

$$(b + (-a)) + a = b + ((-a) + a) \quad \text{from associativity}$$
$$= b + 0 \quad \text{definition of inverse}$$
$$= b \quad \text{definition of identity}$$

Similarly

$$a + ((-a) + b) = (a + (-a)) + b = 0 + b = b$$

Thus

$$b+(-a)$$

is a solution of

$$x+a=b$$

and

$$(-a)+b$$

is a solution of

$$a+y=b$$

To show that these solutions are unique

if $x+a=b$ then

$x=x+0$	definition of identity
$=x+(a+(-a))$	definition of inverse
$=(x+a)+(-a)$	associativity
$=b+(-a)$	hypothesis

hence $x=b+(-a)$

Similarly

$$a+y=b \Rightarrow y=(-a)+b \qquad \qquad \square$$

Discussion

This is an important theorem because it means we can do a variety of familiar manipulations to two equal expressions and deduce that the resulting expressions are still equal. In general it only works with groups however. If we take $\mathscr{P}\{a,b,c\}$ with $\cup$, a monoid, we cannot deduce

$$A=B$$

from

$$A \cup C = B \cup C$$

for example. For instance, take

$$A=\{a\}$$
$$B=\{a,b\}$$
$$C=\{b\}$$

then

$$A \cup C = \{a,b\}$$
$$B \cup C = \{a,b\}$$

but

$$A \neq B$$

Theorem 8.4

In the definition of a group, the identity and inverse conditions can be replaced by weaker conditions

Left-identity: for some i, $i+a=a$ for all a
Left-inverse: given any a, there is another element $-a$ such that $(-a)+a=i$

Proof
With these weaker conditions, cancellation on the left is possible, that is

$$c+a=c+b \Rightarrow a=b$$

for

$$c+a=c+b \Rightarrow$$
$$-c+(c+a)= -c+(c+b) \Rightarrow$$
$$((-c)+c)+a=((-c)+c)+b \Rightarrow \qquad \text{(associativity)}$$
$$i+a=i+b$$
$$a=b$$

Secondly, the given left-identity is also a right-identity

$$-a+a+i=i+i=i= -a+a$$

so by left cancellation of $-a$, that is we add $-(-a)$ to both sides on the left

$$-(-a)+(-a)+a+i= -(-a)+(-a)+a \Rightarrow$$
$$i+a+i=i+a \Rightarrow$$
$$a+i=a \text{ so } i \text{ is also a right identity}$$

Finally, left inverses are also right inverses

$$-a+(a+(-a))=(-a+a)+(-a) \qquad \text{associativity}$$
$$=i+(-a)$$
$$= -a$$
$$= -a+i$$

We can now cancel $-a$ on the left, and obtain

$$a+(-a)=i$$

so $-a$ is a right inverse of a.   □

Corollary

By a similar argument the inverse and identity conditions of a group can be replaced by weaker right inverse and right identity conditions.

Discussion

The theorem does not necessarily hold for identities in monoids. We could, for example, have a monoid in which

$$a+b=b$$

for all a and b. This would be an associative operation because, for all a, b, c

$$a+(b+c)=a+c=c$$

and

$$(a+b)+c=c$$

Every element is a left identity, but if there is more than one element, no element is a right identity

$$a+b=b$$

for $a=b$, so b is not a right identity.

Theorem 8.5

The bijections on a set S with functional composition as the operator form a group.

Proof

The identity element is the identity mapping which maps all $a \in S$ to a, and the inverse of any mapping is the inverse mapping.   □

Example 8.6

(i)   Hence the set of all permutations of a finite set S is a group with $\circ$ as the operator. $\circ$ is associative

$$l \circ (m \circ n) = (l \circ m) \circ n$$

for

$$l \circ (m \circ n)(x) = l(m(n(x))) = (l \circ m) \circ n(x)$$

for all $x \in S$.
(ii)   An example of a group of permutations is transformations of symmetry of a geometrical figure.

Consider the rotations and reflections of a square in figure 8.3.

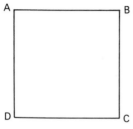

DABC   CDAB   BCDA   ABCD   DCBA   ADCB   BADC   CBAD

Figure 8.3

These are a subset of the permutations of the set $\{A, B, C, D\}$.

The first four (rotations) also form a group which is commutative (defined below).

Definition 8.7

A dyadic operator $*$ defined on a set S is *commutative* (or synonymously, *Abelian* after the mathematicial Abel 1802–1829) if

$$\forall a, b \in S . a * b = b * a$$

A semigroup/monoid/group whose operation is commutative is called a commutative or Abelian semigroup/monoid/group.

Examples 8.7

(i)   The integers or reals with $+$ are groups; with $*$ they are each monoids. The identity for $+$ is 0 and the inverse of a number x is $-x$. The identity for $*$ is 1 but no integer except 1 and -1 has an inverse, and real 0 has no inverse.

(ii)   The integers or reals with $-$ are not semigroups; $-$ is not associative.

(iii)   The reals with $/$ is not a semigroup:—again $/$ is not associative.

(iv)   The integers with $/$ is not a semigroup because not only is $/$ not associative, but the $/$ is not total over the integers.

(v)   The Booleans with $\wedge$ is a monoid: $\wedge$ is associative, True is an identity, but False has no inverse (that is, there is no truth value x such that $x \wedge \text{False} = \text{True}$).

(vi)   The Booleans with $\vee$ is a monoid: $\vee$ is associative, False is an identity, but True has no inverse.

(vii)   The Booleans with $\Rightarrow$ is not a semigroup: $\Rightarrow$ is not associative. Proof: consider the values of

$$\text{False} \Rightarrow (\text{False} \Rightarrow \text{False})$$
$$(\text{False} \Rightarrow \text{False}) \Rightarrow \text{False}$$

(viii) The powerset of a given non-empty set S, $\mathscr{P}(S)$, with $\cup$ is a monoid: $\cup$ is associative, $\{\ \}$ is an identity, but there are no inverses.

(ix) $\mathscr{P}(S)$ with $\cap$ is likewise a monoid.

Definition 8.8
An operation $Op: A \times A \to A$ is said to be *closed* over A if $Op(a, b)$ is defined and a member of A for every pair $(a, b) \in A \times A$.

Example 8.8

(i)   The Reals without zero, $\mathbb{R} - \{0\}$, are closed under $/$, but not the integers (for example, $1/2$ is not a integer).

(ii)   The positive Reals are closed under $\sqrt[+]{\ }$ but not the Rationals.

(iii) For any set S, $\mathscr{P}(S)$ is closed under $\cup$, $\cap$, $-$.

8.6 RINGS AND FIELDS

Definition 8.9
A *ring* is a set A with two operations $+$, $*$, such that

(i) A is a group with respect to $+$
(ii) A is closed with respect to $*$
(iii) $*$ is associative
(iv) $*$ is distributive with respect to $+$; that is, $\forall a, b, c \in A$

$$a*(b*c) = (a*b)*c \ \wedge \qquad \text{(associative)}$$

$$a*(b+c) = (a*b) + (a*c)$$

$$\wedge \qquad \text{(distributive)}$$

$$(a+b)*c = (a*c) + (b*c)$$

Definition 8.10
A ring which is a group with respect to both operations $+$ and $*$, except that 0 has no inverse under $*$, is called a *field*.

Example 8.9

The Reals with respect to $+$ and $*$ are a field: the Reals are a group with respect to $+$ and $*$, for 0 and 1 are respective inverses, because any $r \in \mathbb{R}$, $-r$ is the inverse with respect to $+$ and $1/r$ is the inverse with respect to $*$ (except when $r = 0$).

8.7 OPERATIONS AND ALGEBRAS

Definition 8.11
In general a set with a finite set of operations of various adicities is called
a (*homogeneous*) *algebra*.

Examples 8.10

Semigroups, monoids and groups are examples of sets with dyadic
operations, that is operations with two arguments. $\sim$, $+$ are examples of
operations with one argument, that is monadic operations.

Exercise 8.8

Determine whether the following are rings, fields, or neither

(i)   the integers with respect to $+$ and $*$
(ii) the integers with respect to $+$ and $/$
(iii) the reals with respect to $+$ and $*$
(iv) the reals with respect to $+$ and $/$
(v)   the Booleans with respect to $\wedge$, $\vee$
(vi) the integers modulo n with respect to $+$, $*$. (This is a little harder:
      first show that they form a ring, then consider the cases n prime and n
      not prime.)

Exercise 8.9

Prove that in any ring

$$0*b=0=b*0$$

and that

$$(-a)*(-b)=a*b$$

(*Hint:* $\forall a \in R . a = a + 0$
$\forall a \in R . a*b = (a+0)*b)$

Exercise 8.10

Prove that, in any ring with an identity i with respect to $*$

$$(-i)*a = -a$$

Exercise 8.11

If A and B are two rings, define $+$ and $*$ in $A \times B$ as follows:

$$(a_1, b_1) + (a_2, b_2) = (a_1 + a_2, b_1 + b_2)$$
$$(a_1, b_1) * (a_2, b_2) = (a_1 * a_2, b_1 * b_2)$$

Prove that $A \times B$ with $+$ and $*$ so defined is a ring. If A and B are fields, is $A \times B$ likewise a field?

Examples 8.11

(i)   Semigroups, monoids, groups, rings and fields are all particular kinds of (homogeneous) algebras.
(ii) A finite family of sets closed under $\cap$ and $\cup$ is an algebra.
(iii) The powerset $\mathscr{P}(S)$ of a given set S with $\cup$ and $\cap$ is an algebra.
(iv) A subset of a powerset $\mathscr{P}(S)$ closed under $\cup$ and $\cap$ is an algebra.
(v)   The set of all strings over a given alphabet with the operation of concatenation is an algebra.

8.8 HETEROGENEOUS ALGEBRAS

Examples 8.12

All the algebras we have encountered so far has been homogeneous, which is to say there has been just one set and a number of operations over it. For example, $\mathscr{P}(S)$ with $\cup$ and $\cap$ is a homogeneous algebra, so is the Booleans with $\vee$, $\wedge$, $\sim$, $\Rightarrow$. However, we could usefully combine these so that we can also use the operation $\subseteq$ on $\mathscr{P}(S)$ which gives a result in $\mathbb{B}$. In other words

$$\subseteq : \mathscr{P}(S) \times \mathscr{P}(S) \to \mathbb{B}$$

Another useful operator is the triadic if...then...else... construct, which for any type T is of form

$$\text{if} \ldots \text{then} \ldots \text{else} \ldots : \mathbb{B} \times T \times T \to T$$

If the Boolean expression following the 'if' is true, the value of the total expression is the same as that T-value following the 'then', otherwise it is the same as that T-value following the 'else'.

Definitions 8.12
In general, any algebra is characterised by a *signature* which is a pair $\langle S, \Omega \rangle$. S is a set of types or 'sorts' and Ω is a family of sets of operators. Each operator in a given member of the family has the same type, where the type belongs to $S^* \times S$. This type is often called the *scheme* of the operator. If S contains more than one sort, the algebra is called *heterogeneous* or *many-sorted*.

Examples 8.13

(i) Sets with if... then... else

Given $\mathbb{B}$ and a powerset $\mathscr{P}(S)$, we can construct an algebra with signature $\langle T, \Omega \rangle$ where

$$T = \{S, \mathscr{P}(S), \mathbb{B}\}$$
$$\Omega = \{\{\sim\}: \mathbb{B} \to \mathbb{B}$$
$$\{\vee, \wedge, \Rightarrow\}: \mathbb{B} \times \mathbb{B} \to \mathbb{B}$$
$$\{\cap, \cup, -\}: \mathscr{P}(S) \times \mathscr{P}(S) \to \mathscr{P}(S)$$
$$\{\subseteq\}: \mathscr{P}(S) \times \mathscr{P}(S) \to \mathbb{B}$$
$$\{\text{if}...\text{then}...\text{else}...\}: \mathbb{B} \times \mathscr{P}(S) \times \mathscr{P}(S) \to \mathscr{P}(S)$$
$$\{\text{if}...\text{then}...\text{else}...\}: \mathbb{B} \times \mathbb{B} \times \mathbb{B} \to \mathbb{B}\}$$

Thus in this example there are nine operators, one of which is overloaded. These operators are grouped into a family of six sets, each set with its own scheme.

(ii) Program language procedures

In a high level language a function or procedure without side-effects is an operator of type

$$\underset{i}{\times} S_i \times \underset{j}{\times} S_j$$

where each S_i is the type of each input parameter, and each S_j is the type of each output parameter and any result. The S_i and S_j are drawn from the types used in the program. Thus in a block-structured program written in a language such as ALGOL 60 or Pascal, the main program contains various invocations of operators within an algebra: the sorts of the algebra are the types of all the variables declared in the outermost block and of all the expressions standing as parameters of procedures and functions called from within the outermost block. The operators may be considered to be the procedures and functions called from within the outermost block, with sorts as described. The same ideas can be applied to any block within a program.

In the more recent programming languages such as Ada and CLU, the data type definitions are packaged together with all the procedure definitions which make use of them, giving a clearer association with the algebra which they comprise. For this reason such constructs are sometimes called 'Abstract Data Types' (although many writers, including myself, prefer to reserve this phrase for describing definitions which exclude any algorithmic definition of procedure bodies, thereby divorcing the definition of what a procedure does from how it does it).

(iii) Expressions in programming languages
In any program there is, at any point, the potential for writing a large variety of expressions. The type of any such an expression is a member of

$$S^* \times T$$

where S is the set of the types of all the variables visible at that point (that is, in the scope of whose declarations that point lies), and T is the union of S and any other types which the language allows to be constructed from S. (For most languages T will be simply identical to S, but in ALGOL 68, for example, the result of an expression could be a denotation of type **row** A where A is a visible type. Thus if the language allows explicit type construction mechanisms T may be larger than S).

Note that the above only applies if the expression does not have side-effects. (Thus we already see that a mathematical statement of the effects of a computer program can only be made if the rules of structured programming are followed.) For example, consider

> **int** X, Y;
>
> **real** e, r;
>
> **boolean** t, s;
>
> **proc** PX (a, b, c, d);
>
> >**int** a, b; **real** c, d;
> >
> >IN a, c; OUT d; INOUT b;
> >
> >**begin**
> >
> >$\vdots$
> >
> >**end**;
>
> $\vdots$
>
> $X := $ **if** $e/r = 3.4$ **or** t **and** s **then** $X + 1$ **else** Y;
>
> $\vdots$
>
> $\ldots (e + r) / 1.4159 \ldots$

The procedure PX invokes a function

$$f_{PX} : \mathbb{Z} \times \mathbb{Z} \times \mathbb{Z} \to \mathbb{Z} \times \mathbb{R}$$

and the expression on the right-hand side of the assignment expresses a function

$$\mathbb{R} \times \mathbb{R} \times \mathbb{B} \times \mathbb{B} \times \mathbb{Z} \times \mathbb{Z} \to \mathbb{Z}$$

The last expression expresses a function

$$\mathbb{R} \times \mathbb{R} \to \mathbb{R}$$

(iv) Personnel file

In our example of a personnel file the file of employees represented data of type

$$Employees = Employee\text{-}Id \longmapsto (Names \times Ages \times Grades \times Salaries)$$

where the $\longmapsto$ symbol is understood to represent a partial mapping. One would expect to provide various operations on this file, namely

Get details of an employee	(GET)
Add a new entry	(ADD)
Delete an entry	(DELETE)
Change an entry	

the last operation would come in four variations

Change name	(CHNAME)
Change age	(CHAGE)
Change grade	(CHGRADE)
Change salary	(CHSALARY)

These operations together with the sets involved form an algebra with signature $\langle T, \Omega \rangle$ where

$T = \{Employees,\ Employee\text{-}Id,\ Names,\ Ages,\ Grades,\ Salaries\}$

$\Omega = \{\{DELETE\}$: $Employees \times Employee\text{-}Id \rightarrow Employees$

$\quad \{ADD\}$: $Employee \times Employee\text{-}Id \times Names \times Ages$
$\qquad\qquad\qquad \times Grades \times Salaries \rightarrow Employees$

$\quad \{GET\}$: $Employees \times Employee\text{-}Id \rightarrow Names \times Ages$
$\qquad\qquad\qquad \times Grades \times Salaries$

$\quad \{CHNAME\}$: $Employees \times Employee\text{-}Id \times Names \rightarrow Employees$

$\quad \{CHAGE\}$: $Employees \times Employee\text{-}Id \times Ages \rightarrow Employees$

$\quad \{CHGRADE\}$: $Employees \times Employee\text{-}Id \times Grades \rightarrow Employees$

$\quad \{CHSALARY\}$: $Employees \times Employee\text{-}Id \times Salaries$
$\qquad\qquad\qquad \rightarrow Employees\}$

When we write procedures to implement these operations, the types of the parameters of these procedures should, in some perhaps indirect way, reflect the sorts of the corresponding operators.

Also, of course, we have not yet stated what these operators actually do, only the types of their arguments and results. We shall see how to specify the 'semantics', that is the effect of the operation, in chapter 9.

(v) Transactions on a file

Returning to our transactions on a file example, first introduced in chapter 2, we had four operations

> *Add*
> *Update*
> *Delete*
> *Purge*

and our data types were

> *File*
> *R*
> *INF*

We can now suggest a signature for this total algebra as $\langle T, \Omega \rangle$ where

> $T = \{File, R, INF\}$
> $\Omega = \{\{Purge\}: File \rightarrow File$
> $\quad \{Delete, Add\}: File \times R \rightarrow File$
> $\quad \{Update\}: File \times R \times INF \rightarrow File\}$

(vi) Stacks

A familiar and, perhaps these days, laboured example of an abstract programming construct is a stack. If the types of items which are placed on the stack is *T*, and *Stack* denotes the type which is a stack, then the signature of an algebra for the stack and its operations is $\langle S, OP \rangle$ where

> $S = \{Stack, T, \{empty\}\}$
> $OP = \{\{Push\}: Stack \times T \rightarrow Stack$
> $\quad \{Pop\}: Stack \rightarrow Stack$
> $\quad \{Top\}: Stack \rightarrow T \cup \{empty\}\}$

In this collection of operators we have chosen to make *Top* deliver the topmost item from the stack without changing the stack and the message 'empty' if the stack is empty, and *Pop* merely removes the topmost item from the stack without giving it back. There are a number of equivalent variations.

(vii) Arrays

Special procedures may be written in a high level language for operating on arrays, to provide array addition, subtraction, multiplication, inversion etc. The resulting array algebra would have a suitable signature.

Exercise 8.12 (Workshop)

In the space-borne computer problem (see exercises 3.10 and 6.6) the data types were

> *position*
> *velocity*

acceleration
mass
burn-rate

etc. Various operations can now be defined: for example, the position of the craft at some point in the future can be calculated from the present position, the velocity and the time ahead (assuming no acceleration). This can be represented by an operator with a suitable scheme. A similar operator can be defined to predict the velocity given an acceleration, and the acceleration given the heavenly bodies, their masses, the burn rate and mass of the craft. Give an algebra with its signature including these operators and their schemes, together with any other useful-seeming operators that may occur to you.

Exercise 8.13 (Workshop)

In the fault-tolerant control system (see exercises 6.7 and 3.11) treat the checkers and arbiter as operators (assume all checkers invoke the same operator) and define their schemes. Define an operator for the whole system which takes as its input arguments the input data d in D and the set of functional blocks. (The types of these will also have to be defined.) Then define a signature for the whole system.

Exercise 8.14 (Workshop)

In the personal diary problem (see exercises 6.8, 4.5 and 3.9), devise schemes for the operations of adding, deleting and moving an event in the diary, and give the signature of the resulting algebra.

Exercise 8.15 (Workshop)

In the nuclear power station exercise (see exercises 6.9, 4.4 and 3.8) the action of the program in sending values to the dials and lamps in response to the sensors, and in controlling the positions of the rods and displaying messages, can be regarded as a number of algebraic operators. Extend the work done in exercise 6.9 by defining the schemes of these operators.

8.9 FURTHER READING

A general study of algebra is very rewarding, and to pursue the subject further MacLane and Birkhoff (1979) *Algebra* (Second Edition) is greatly recommended.

SUMMARY OF RESULTS

Result	Reference
For a dyadic operation there is at most one identity element	Theorem 8.1
String concatenation is associative	Theorem 8.2
Cancellation theorem for groups	Theorem 8.3
In the definition of a group, the identity and inverse conditions can be replaced by weaker left identity and left inverse conditions	Theorem 8.4
Bijections on a set with functional composition form a group	Theorem 8.5

Chapter 9 Homomorphisms

NOTATION INTRODUCED

Notation	Concept	Reference
	homomorphism	Def. 9.1
	epimorphism	Def. 9.2
	monomorphism	Def. 9.3
	isomorphism	Def. 9.4
	homomorphisms of general algebras	Def. 9.5
$h(|A|), h(A)$	range of a homomorphism	Section 9.4
A/h	quotient algebra	Def. 9.6
	kernel of a homomorphism	Def. 9.7
	endomorphism	Def. 9.8
	automorphism	Def. 9.9
h^n	exponentiation of endomorphisms	Section 9.7
	free, freely generated groups	Def. 9.10
	adequate representation of data	Def. 9.11
	adequate representation of operations	Def. 9.12
	correct satisfaction of requirements	Def. 9.13

In chapter 6 we saw how mappings can relate sets to each other, and we defined various properties of mappings such as 'one–one', 'total', 'partial' and so forth. Then in chapter 8 we were introduced to algebras, which are sets, or collections of sets, together with operations. This chapter introduces the idea of a homomorphism, which is a mapping extended to algebras: in other words a homomorphism is a mapping which relates the sets underlying two algebras and which preserves their 'structure', that is the operations defined upon them.

Whenever one designs data in a program, and procedures to operate on that data, one is constructing homomorphisms. For example, one may use files of names to represent a set of names; then various file manipulations may be devised to represent the operations of union and intersection, and the membership relation. Records in programming languages are frequently used to represent Cartesian products. In these cases there are mappings from files to sets and from records to products which respect or preserve the operations. We shall start by introducing homomorphisms as mathematical ideas and then later in the chapter show some examples in the world of programming.

9.1 BASIC DEFINITIONS

Definition 9.1
If an operation '+' is defined on two separate sets A and B, then a mapping $h: A \rightarrow B$ is a *homomorphism* if, for all $a, b \in A$

$$h(a + b) = h(a) + h(b)$$

The definition can be extended for algebras with more than one operation or more than one set.

A homomorphism will preserve various properties of an algebra. For example if A and B are both groups, h will map identities to identities and inverses to inverses.

Theorem 9.1

If h is a homomorphism mapping $A, +$ to $B, +$ and 0 is an identity for A, then $h(0)$ is an identity for B.

Proof

$$\forall a \in A . 0 + a = a \qquad \text{—definition of identity}$$
$$\forall a \in A . h(0 + a) = h(0) + h(a)$$
$$= h(a) \qquad \text{—definition of homomorphism}$$
$$h(0) \text{ is an identity for } B \qquad \qquad \square$$

By a similar argument, a homomorphism maps inverses to inverses.

Theorem 9.2

If h is a homomorphism mapping $A, +$ to $B, +$ and 0 is an identity for A, then for all $a \in A$ for which the inverse $-a$ exists, $h(-a) = -h(a)$

Proof

$$\forall a \in A \,.\, h(a) + h(-a) = h(a + (-a)) = h(0)$$

But $h(0)$ is the identity for B.

Therefore $h(-a)$ is the inverse of $h(a)$. $\quad\square$

Homomorphisms may exist between two semigroups, two monoids, two rings, or any other algebraic structures. For rings, which have two operations $*$ and $+$ say, we would require that, for all a and b

$$h(a) + h(b) = h(a + b)$$
$$h(a) * h(b) = h(a * b)$$

Rules, like associativity or commutativity, are preserved by homomorphisms. Let us take the associative rule.

Theorem 9.3

Given a homomorphism

$$h: A, + \rightarrow B, +$$

where the associative rule holds in A. Then

$$\forall a, b, c \in A \,.\, h(a) + (h(b) + h(c)) = (h(a) + h(b)) + h(c)$$

Proof

$$\forall a, b, c \in A \,.\, a + (b + c) = (a + b) = c$$

—associativity

Therefore

$$h(a + (b + c)) = h((a + b) + c)$$

—homomorphism

But

$$h(a + (b + c)) = h(a) + h(b + c)$$
$$= h(a) + (h(b) + h(c))$$

and

$$h((a + b) + c) = h(a + b) + h(c)$$
$$= (h(a) + h(b)) + h(c)$$

so

$$h(a) + (h(b) + h(c)) = (h(a) + h(b)) + h(c)$$

$\quad\square$

Exercise 9.1

Show that, if h is a homomorphism mapping $A,*$ to $B,*$, then if $*$ with A is commutative, so is $*$ with B.

Exercise 9.2

If g maps a ring R to a ring S, show that g preserves the distributive law.

9.2 TYPES OF HOMOMORPHISM

Homomorphisms with various different properties are given different names. Most of these properties relate to the kinds of mapping inherent in the homomorphism.

Definition 9.2
If a homomorphism $h:A \rightarrow B$ maps A onto B—that is, if every element of B is an image of some element of A under h, then h is said to be an *epimorphism.*

Example 9.1

For example, the group consisting of the integers and $+$ can be mapped by a homomorphism f to the group consisting of the integers modulo n and $+$: for any integer x, $f(x) = x \bmod n$. Furthermore, every element of the integers modulo n (that is, the numbers $0, 1, \ldots, n-1$) is an image of some integer under f (for example, $0, 1, \ldots, n-1$, or $n, n+1, \ldots, 2n-1$). So in this case f is an epimorphism.

Exercise 9.3

Show in the above example that f preserves $+$.

Example 9.2

For another example, we can take collections of sets which are closed under intersection and union. Consider the families of sets

$$A = \{\{ \ \}, \{a\}, \{a,b\}\}$$

and

$$B = \{\{ \ \}\}$$

We define two homomorphisms: $h:A \rightarrow B$ and $g:B \rightarrow A$. h maps every member of A to the unique member $\{ \ \}$ of B. g maps $\{ \ \}$ in B to $\{ \ \}$ in A.

Now h is an epimorphism because all the members of B are the image of some member of A under h. However, g is not an epimorphism because there are members of A ($\{a\}, \{a,b\}$) which are not images of some member of B under g.

Exercise 9.4

Show in the above example that both g and h preserve the operations of union and intersection.

Definition 9.3
A homomorphism which is an injection—that is, one to one—is called a *monomorphism*.

Example 9.3

In example 9.2 above, g is a monomorphism but f and h are not.

Definition 9.4
A homomorphism which is both an epimorphism and a monomorphism is called an *isomorphism*.
 'Isomorphism' means 'having the same shape'. If two structures A and B are such that there is an isomorphism between them, they are essentially 'the same'. The only difference between them is the names of their elements. They are called isomorphic.

Example 9.4

Consider the set of sets

$$C = \{\{\ \}, \{x\}\}$$

with union and intersection. Then union can be expressed by the table

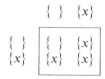

and intersection by

	$\{\ \}$	$\{x\}$
$\{\ \}$	$\{\ \}$	$\{\ \}$
$\{x\}$	$\{\ \}$	$\{x\}$

Consider now the familiar Boolean algebra $\{T, F\}$ with $\wedge$ and $\vee$. These operations can be defined by the tables

$\wedge$	T	F
T	T	F
F	F	F

$\vee$	T	F
T	T	T
F	T	F

A homomorphism m mapping $T \mapsto \{\ \}$ and $F \mapsto \{x\}$ now preserves the operations: $\wedge$ is mapped to union and $\vee$ to intersection. (Check this by inspecting the tables.) Since m is one-to-one and onto, it is an isomorphism.

As with bijections, isomorphisms have inverses which are also isomorphisms. In this example m^{-1} maps $\{\ \} \mapsto T$, $\{x\} \mapsto F$ and maps union to $\wedge$ and intersection to $\vee$.

There is another pair of isomorphisms in this example, mapping $T \mapsto \{x\}$, $F \mapsto \{\ \}$, $\wedge$ to intersection and $\vee$ to union.

Exercises 9.5

Consider the following families of sets, each closed under the operations of union and intersection

$$A = \{\{\ \}\}$$
$$B = \{\{\ \}, \{a\}\}$$
$$C = \{\{\ \}, \{b\}\}$$
$$D = \{\{\ \}, \{a\}, \{a, b\}\}$$
$$E = \{\{\ \}, \{b\}, \{a, b\}\}$$
$$F = \{\{\ \}, \{a\}, \{b\}, \{a, b\}\}$$

There are homomorphisms between any pairs of these which preserve $\cup$ and $\cap$. Find examples of epimorphisms, monomorphisms, and isomorphisms. Is there any situation in which more than one homomorphism exists mapping one family to another?

9.3 HOMOMORPHISMS OF HETEROGENEOUS ALGEBRAS

(This section may be omitted at a first reading.)

Recall from chapter 8.8 that in general an algebra is characterised by a signature

$$\Sigma = \langle S, \Omega \rangle$$

where S is a set of sorts and Ω a family of sets of operators. We can extend the definition of homomorphism to cater for heterogeneous algebras as follows.

Definition 9.5

A homomorphism h from an algebra A with signature $\Sigma_A = \langle S_A, \Omega_A \rangle$ to an algebra B with signature $\Sigma_B = \langle S_B, \Omega_B \rangle$ maps each sort s in S_A to a sort $h(s)$ in S_B, and each operator p in each set in Ω_A to an operator $h(p)$ in one of the sets in Ω_B such that

(i) For all the operators p in A, that is

$$\forall p \in \bigcup_{P \in \Omega_A} P$$

the arities of p and $h(p)$ are equal.

(ii) If p is an operator in the algebra A, then let its type be

$$S_{p,1} \times S_{p,2} \ldots \times S_{p,n} \rightarrow S'_p$$

where n is the arity of the operator, and

$$S_{p,1}, S_{p,2}, \ldots, S_{p,n}, S'_p \in S_A$$

Then

$$\forall x_1 \in S_{p,1}, x_2 \in S_{p,2}, \ldots, x_n \in S_{p,n} . h(p(x_1, x_2, \ldots, x_n))$$
$$= h(p)(h(x_1), h(x_2), \ldots, h(x_n))$$

The formal notation for this looks more hair-raising than it really is. It is a simple extension from an algebra, such as a group, involving one set and one operator. In fact we have already seen an example of a homomorphism of heterogeneous algebras.

Example 9.5

We saw there was a homomorphism m from

$$\{T, F\}, \{\wedge, \vee\}$$

to

$$\{\{\ \}, \{x\}\}, \{\cup, \cap\}$$

Strictly, to define the homomorphism m we would need to declare

$m(T) = \{\ \}$

$m(F) = \{x\}$

$m(\wedge) = \cup$

$m(\vee) = \cap$

These algebras are already more elaborate than a group because they have two operators. However each has only one underlying set, or carrier as it is called: the first has $\{T, F\}$ as a carrier, the second $\{\{\ \}, \{x\}\}$.

An important point about homomorphisms is that, given any new operator defined on the domain algebra of the homomorphism, another operator is induced on the range algebra. In the above example, we could further define $\sim$

$\sim T = F$

$\sim F = T$

This would induce an operator on the range algebra $m(\sim)$ as follows

$m(\sim)(m(T)) = m(\sim)(\{\ \})$

$\qquad = m(\sim T) = m(F) = \{x\}$

and

$m(\sim)(m(F)) = m(\sim)(\{x\})$

$\qquad = m(\sim F) = m(T) = \{\ \}$

So the table for the operator $m(\sim)$ is

	x	
	$\{\ \}$	$\{x\}$
$m(\sim)(x)$	$\{x\}$	$\{\ \}$

which is the complete operation.

Exercise 9.6

What is the operator on $\{\{\ \}, \{x\}\}$ induced by m on $\Rightarrow$?

Exercise 9.7

Taking the other homomorphism (actually the more familiar one) mapping $F \mapsto \{\ \}$, $T \mapsto \{x\}$, $\wedge \mapsto \cap$, $\vee \mapsto \cup$, what operators in $\{\{\ \}, \{x\}\}$ are induced by $\sim$ and $\Rightarrow$?

9.4 QUOTIENT ALGEBRAS

(These next two sections may be omitted at a first reading.)

Associated with homomorphisms is the idea of a Quotient Algebra. We shall introduce this idea in stages. First, some notation and then a theorem.

Notation

If A and B are algebras with single-set carriers, their carriers are denoted by $|A|$ and $|B|$. If

$$h: A \to B$$

is a homomorphism, its range is denoted by $h(|A|)$.

Theorem 9.4

Given $h: A \to B$ where A and B are algebras with single-set carriers, then $h(|A|)$ is closed under the operators of $h(A)$.

Proof

Let p be any operator in A (more strictly, let the signature of A be

$$\Sigma_A = \langle S_A, \Omega_A \rangle$$

and let

$$p \in \bigcup_{P \in \Omega_A} P$$

Let the arity of p be n. Then

$$h(|A|) = \{x | x \in |B| \wedge (\exists a \in |A| \,.\, x = h(a))\}$$

Therefore, given any $b \in h(|A|)$ we can find $a \in |A|$ such that $b = h(a)$. So, given $b_1, b_2, \ldots, b_n \in h(|A|)$ we can find $a_1, a_2, \ldots, a_n \in |A|$ where $b_1 = h(a_1), b_2 = h(a_2), \ldots, b_n = h(a_n)$.

Now, since p is closed in A

$$\forall a_1, a_2, \ldots, a_n \in |A| \,.\, p(a_1, a_2, \ldots, a_n) \in |A|$$

and so

$$\forall a_1, a_2, \ldots, a_n \in |A| \,.\, h(p(a_1, a_2, \ldots, a_n)) \in h(|A|)$$

However, by the rules of homomorphisms

$$h(p(a_1, a_2, \ldots, a_n)) = h(p)(h(a_1), h(a_2), \ldots, h(a_n)) \in h(|A|)$$

and so

$$\forall b_1, b_2, \ldots, b_n \in h(\|A\|) . h(p)(b_1, b_2, \ldots, b_n) \in h(\|A\|)$$

so that $h(p)$ is closed in $h(\|A\|)$ and hence in $\|B\|$. This will hold for all operators in A. $\square$

Thus, in general terms, the range of a homomorphism, consisting of the images of the carrier sets with the images of the operations form a closed algebra.

Notation

Because of the above theorem, the image of an algebra A under a homomorphism h is simply written $h(A)$.

Example 9.6

Looking back at the example in exercise 9.5, we can define a homomorphism $k: E \rightarrow D$ as shown in figure 9.1. Then

$$k(\|E\|) = \{\{ \ \}, \{a\}\}$$
$$k(E) = \langle \{\{ \ \}, \{a\}\}, \{\cap, \cup\} \rangle$$

and $k(E)$ is clearly a closed algebra.

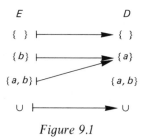

Figure 9.1

Notation

If we take the domain algebra of a homomorphism h, we can partition its carrier sets into non-intersecting subsets. Each carrier set S can be partitioned as follows

$$Part_h(S) = \{P \mid P \subseteq S \wedge (\forall a, b \in P . h(a) = h(b))\}$$

Thus in the example above

$$Part_k(E) = \{\{\{ \ \}\}, \{\{b\}, \{a, b\}\}\}$$

Theorem 9.5

Given $h:A \to B$ where S is a carrier of A, $Part_h S$ is in one–one correspondence with $h(S)$.

Proof
Let

$$m:h(S) \to Part_h(S)$$

be defined as follows

$$m(x) = \{ y | y \in S \wedge h(y) = x \}$$

This is clearly a member of $Part_h(S)$ by definition of the latter. Furthermore, it we take a member P of $Part_h(S)$, by convention members of a partition are not empty, and for each $y \in P$, $h(y)$ is identical. Hence m^{-1} is defined and

$$m^{-1}(P) = h(y)$$

for any y in P. m and m^{-1} are clearly inverses. The theorem follows. □

Example 9.7

The above theorem may be made clearer by illustrating with our previous example. We can further define the original operators of E on $Part_k(E)$ so that

$$\forall x, y \in Part_k(E) . x \cap y \in Part_k(E) \wedge$$

$$(\forall q \in x \cap y, r \in x, s \in y . h(q) = h(r)h(\cap)h(s))$$

This is a bit obscure: it is perhaps best illustrated with a diagram. Consider figure 9.2. $k(E)$ now induces subsets of $|E|$ 'reflecting back' through the homomorphism (figure 9.3). $Part_k(E)$ now has two elements, just as $k(|E|)$ does, and so we can define isomorphic operators $\cap$, $\cup$ on it (figure 9.4).

It should be now intuitively clear that $Part_k(E)$ with the operators $\cap$ and $\cup$ is isomorphic to $k(E)$. In fact $Part_k(E)$ is called the quotient algebra and is written E/k.

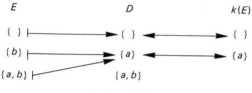

Figure 9.2

Part$_k$(E)   k(E)

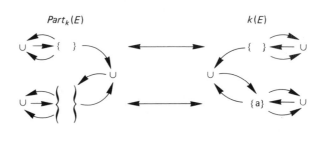

Figure 9.3

Part$_k$(E)   k(E)

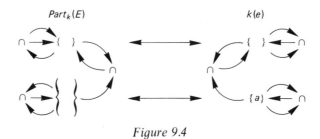

Part$_k$(E)   k(e)

Figure 9.4

Definition 9.6
Given $h:A \rightarrow B$, $Part_k(A)$ with the operators extended to $Part_k(A)$ is called the *quotient algebra* and is written A/h.

The quotient algebra is always isomorphic to the range algebra restricted to the range of the homomorphism.

9.5 QUOTIENT GROUPS

If we have a homomorphism h from a group F to a group G, the quotient group G/h has particular properties.

Theorem 9.6

If $h:F \rightarrow G$ where F and G are groups whose identities are 0_F and 0_G respectively, then the subset of F which is mapped to 0_G is closed under the operation of F and is a sub-group.

Proof

First of all, we know that

$$h(0_F)=0_G$$

Secondly, if

$$h(a)=0_G$$

then

$$h(-a)=0_G$$

because

$$0_G=h(0_F)=h(a+(-a))$$
$$=h(a)+h(-a)$$
$$=0_G+h(-a)$$
$$=h(-a)$$

Furthermore, if

$$h(a)=0_G$$

and

$$h(b)=0_G$$

then

$$h(a+b)=h(a)+h(b)=0_G$$

Thus the subset of F which is mapped to 0_G by h is closed under the operation of F. Since it includes the identity and inverses of all the elements within it, it is a sub-group of F. $\square$

Definition 9.7

If $h:F\rightarrow G$ is a group homomorphism, the sub-group of F whose members are mapped to 0_G is called the *kernel* of the homomorphism h. This is illustrated in figure 9.5.

Theorem 9.7

Given a group homomorphism $h:F\rightarrow G$, if $a\in F$ and p is a member of the kernel of h, then $h(a)=h(p+a)=h(a+p)$.

Exercise 9.8

Prove theorem 9.7.

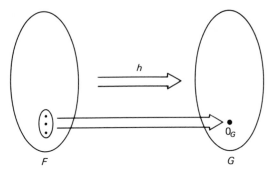

Figure 9.5

Theorem 9.8

The converse is also true; that is, given a group homomorphism $h:F\rightarrow G$, if $a, b \in F$ such that $h(a)=h(b)$ then $b=p+a$ for some member p of the kernel of h.

Proof

$$p=b+(-a)$$

for then

$$\begin{aligned}p+a&=(b+(-a))+a\\&=b+((-a)+a)\\&=b+0_F=b\end{aligned}$$

Secondly

$$\begin{aligned}h(p)&=h(b+(-a))\\&=h(b)+h(-a)\\&=h(b)-h(a)\end{aligned}$$

because homomorphisms preserve inverses

$$=h(a)-h(a)$$

because $h(b)=h(a)$ by hypothesis

$$=0_G$$

Therefore p is in the kernel of h.   □

Corollary 1
Thus if we take any member a of F and add to it all the different members of the kernel, we get that subset of F whose members map to $h(a)$.

Corollary 2
The same argument applies to operation (addition) on the right; that is, in the proof above we could show that b is expressible as $a + q$ where $q = (-a) + b$.

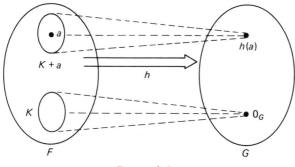

Figure 9.6

9.6 MORE TYPES OF HOMOMORPHISM

Definition 9.8
A homomorphism may have the same algebra as its domain and its range. Such a homomorphism is called an *endomorphism*.

Example 9.8

If we take the integers with $+$, the mapping

$$x \mapsto n * x$$

where n is some integer, is an endomorphism. The integers of the form $n * x$ with $+$ are a group, and a sub-group of the integers.

If n is zero we have all the integers mapped to 0. This is still a group with $+$, but a rather degenerate one.

Example 9.9

'Projection', 'Rotation', 'Translation' are common forms of endomorphism in co-ordinate geometry. If we have the set of two-dimensional vectors on a real plane

$$\mathbb{R} \times \mathbb{R}$$

then vector addition is defined

$$\langle x, y \rangle + \langle p, q \rangle = \langle x + p, y + q \rangle$$

with

$$\langle 0,0 \rangle$$

as identity and

$$\langle -x, -y \rangle$$

as the inverse of $\langle x, y \rangle$

This set of two-dimensional vectors is a group with the operation of $+$. (Exercise: show associativity.)

Any linear mapping

$$\langle x, y \rangle \mapsto \langle ax + by, Ax + By \rangle$$

gives us a homomorphism. To prove it we have to show that, if m is the homomorphism

$$m(\langle x, y \rangle + \langle p, q \rangle) = m(\langle x, y \rangle) + m(\langle p, q \rangle)$$

Exercise 9.9

Prove the above.

Definition 9.9
If an endomorphism is an isomorphism (that is, one-to-one and onto) it is called an automorphism.

Exercise 9.10

In the example used in exercise 9.9 find values of a, b, A, B which make m an isomorphism and other values which make m not an isomorphism.

Example 9.10

A trivial example of an automorphism is the identity mapping of an algebra to itself.

9.7 COMPOSITION OF HOMOMORPHISMS

Homomorphisms compose to form further homomorphisms, in the same way as mappings do.

Theorem 9.9

If

$$h: A \rightarrow B$$

and

$$k:B \to C$$

are homomorphisms, then

$$k \circ h:A \to C$$

is a homomorphism from A to C such that

$$k \circ h(a) = k(h(a))$$

Proof

We give the proof for group homomorphisms.

If A, B, C are groups, then to show that, given h and k are homomorphisms, $k \circ h$ is a homomorphism, we need to show that for all $a, b \in A$

$$k \circ h(a + b) = k \circ h(a) + k \circ h(b)$$

The proof is straightforward

$$
\begin{aligned}
k \circ h(a + b) &= k(h)(a + b) \\
&= k(h(a) + h(b)) &&\text{since } h \text{ is a homomorphism} \\
&= k(h(a)) + k(h(b)) &&\text{since } k \text{ is a homomorphism} \\
&= k \circ h(a) + k \circ h(b) &&\qquad\qquad\square
\end{aligned}
$$

Notation

If $p, q: A \to A$ are endomorphisms, then clearly $p \circ q: A \to A$ is also an endomorphism. We can then define the notation

$$p^2$$

to mean

$$p \circ p$$

and in general

$$p^n \triangleq p \circ p^{n-1}$$
$$p^1 \triangleq p$$

and sometimes it may be helpful to use

$$p^0 = I$$

where I is the identity endomorphism (automorphism) defined

$$I(a) = a$$

for all $a \in A$.

Exercise 9.11

(i)   If h, k are epimorphisms, show that $k \circ h$ is an epimorphism.
(ii) If h, k are monomorphisms, show that $k \circ h$ is a monomorphism.
(iii) If h, k are isomorphisms, show that $k \circ h$ is an isomorphism.
(iv) If h is an isomorphism and k an epimorphism, what is $k \circ h$?
(v)   If h is an epimorphism and k an isomorphism, what is $k \circ h$?
(vi) What are the answers to (iv) and (v) if 'epimorphism' is replaced by 'monomorphism'?

9.8 EXAMPLES

Example 9.11: Transactions on a File

A file may represent a collection of information which can be updated, amended, deleted, etc. Each of the operations (update, delete, etc.) will cause the file to change its 'state', and certain sequences of operations (add item A, add item B, delete item A, delete item B) may cause the file to return to the same state.

We may consider all possible sequences of operations on the file, including the empty sequence, which will have no effect on the state of the file. These sequences, with concatenation as their operator, form a monoid—the empty sequence being the identity element. The sequences of operations on the file cause the file to transit from one state to another, and we can identify those sequences which cause the same transition.

Thus if the 'atomic' operations on the file are $A, B, C, \ldots$, the possible state transitions of the file can be represented by sequences

> A
> AA
> BA
> AB
> ACA
> $\vdots$

The empty sequence may be represented by $\varnothing$. Furthermore, some operations may have inverses—for example, $\bar{A}, \bar{B}$. (For example, 'add an item x', 'delete an item x'). Then $A\bar{A}$ would have the same effect as $\varnothing$. Further, if $\bar{A}$ is the inverse of A, $\bar{B}$ of B etc., then

> $\bar{A}\bar{B}$ is the inverse of BA
> $\bar{A}\bar{A}$ is the inverse of AA
> $BA\bar{C}$ is the inverse of $C\bar{A}\bar{B}$

etc. We would define equality on these sequences such that two sequences are equal if they cause the same transition on the files

$$A\bar{A} = \bar{A}A = B\bar{B} = \bar{B}B = \emptyset$$
$$\emptyset A = A\emptyset = A$$

There may be some equalities not deducible from inverses—for example, possibly

$$BAC = D$$

or

$$DA = AD$$

In particular, if every element of the monoid (that is, operation on the file) has an inverse, the sequences of operations form a group.

Example 9.12: State Machines

In many engineering systems, especially in digital electronics and telecommunications, designers like to think of the systems they are designing as state machines of some kind. A state machine is similar in principle to the file with its transactions. The machine may be capable of adopting a number of different states, and there may be a number of operations on the machine causing it to switch from one state to another. Thus again one may compose sequences of operations which form a semigroup, or a monoid if one includes the empty sequence.

There is a familiar state machine which has six basic operations

$$\{R, L, U, D, F, B\}$$

and all these have the property that

$$R^4 = L^4 = U^4 = D^4 = F^4 = B^4 = 1$$

where 1 is the identity. This means that $R^3 = \bar{R}$, $R = \bar{R}^3$ and $R^2 = \overline{(R^2)}$, and similarly for the other operations. Also there is a limited commutativity, in that

$$RL = LR$$
$$DU = UD$$
$$FB = BF$$

but

$$RU \neq UR$$
$$RD \neq DR$$

etc. Many other more complex relations hold. The machine has a very large number of states, of the order of 4.3×10^{19}, and is the familiar toy, Rubik's cube. If one holds the cube so that the axes through the faces are stationary, R represents turning the right-hand face through 90°, L, U, D, F and B turning the left, upper, downwards-facing, front and back faces respectively.

This state machine forms a group: every sequence of moves has an inverse (otherwise the puzzle would not be solvable!). In general, the alphabet $\{R, L, U, D, F, B\}$ is called the set of generators of the group, because all the elements of the group can be generated by sequences of this alphabet.

Definition 9.10
A group is said to be *freely generated* by its generators if all combinations of the generators (combined using the group's operation) are different. Thus, the group freely generated by the alphabet $\{R, L, U, D, F, B\}$ would be the set of all strings R, RL, RR, RU, RURDRRDDLL, etc. together with their inverses. Such a group is said to be *free*. (The strict algebraic definition of a free group is a little different from this, but the definition given here will serve us.)

Example 9.13

If we consider the free group generated by $\{R, L, U, D, F, B\}$, there is a homomorphism from this group to that representing Rubik's cube obtained by mapping the generators in the obvious manner

$$R \mapsto R$$
$$L \mapsto L$$

etc. Then clearly any combination of elements of the free groups will map to the same combination of the Rubik's cube group

$$RL \mapsto RL$$
$$R^2 \mapsto R^2$$
$$R^4 \mapsto R^4 = 1$$

etc. Thus, if a and b are elements of the free group, and h the homomorphism, we have the equation essential to homomorphisms

$$h(a) . h(b) = h(a . b)$$

The target groups, namely the Rubik's cube group, has more equations relating its elements than the free group; for example

$$R_R L_R = L_R R_R$$

but

$$R_F L_F \neq L_F R_F$$

Where R_R, L_R are the elements R and L in the Rubik's cube group and R_F and L_F are the elements R, L in the free group. It is still the case though, that

$$h(R_F . L_F) = R_R . L_R = h(R_F) . h(L_F)$$

and

$$h(L_F . R_F) = L_R . R_R = h(L_F) . h(R_F)$$

It is 'incidental' to the homomorphism that

$$h(L_F . R_F) = h(R_F . L_F)$$

This is a property common to all homomorphisms, namely that all equations on the domain algebra hold when applied to the mapped elements in the range algebra, and that further equations may hold between elements in the range algebra.

Example 9.14

If S is the set of atomic operators in the algebra which represents the transactions on a file, then that algebra is the range of a homomorphism from the algebra freely generated by S in the same way.

It is in fact a general result that any group is the image of a homomorphism from the group freely generated from its own generators.

Exercise 9.12

What is the group freely generated by an alphabet of one character?

Example 9.15: Integers

If we take the integers with $+$, and the homomorphism of the integers to the integers modulo n (say $n = 5$), we have two groups each with a single generator 1. The first is a free group in that any integer can be expressed as

$$1 + 1 + \ldots + 1$$

or

$$(-1) + (-1) + \ldots + (-1)$$

The second identifies $1 + 1 + 1 + 1 + 1$ with 0; each can be regarded as representable by tallies (figure 9.7). Now each group can be seen as

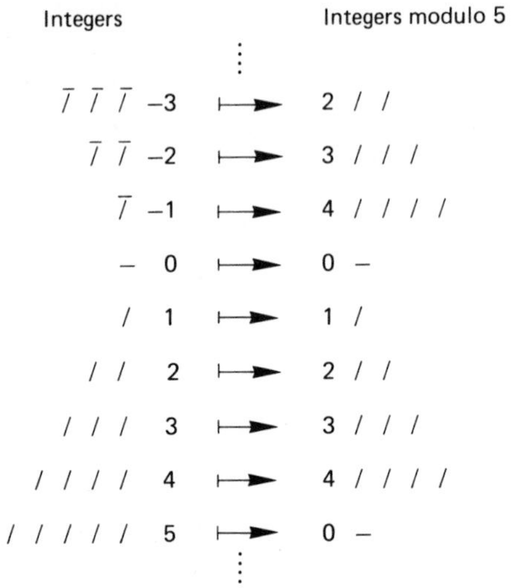

Integers   Integers modulo 5

$\overline{/}\ \overline{/}\ \overline{/}\ $ –3 $\longmapsto$ 2 / /

$\overline{/}\ \overline{/}\ $ –2 $\longmapsto$ 3 / / /

$\overline{/}\ $ –1 $\longmapsto$ 4 / / / /

– 0 $\longmapsto$ 0 –

/ 1 $\longmapsto$ 1 /

/ / 2 $\longmapsto$ 2 / /

/ / / 3 $\longmapsto$ 3 / / /

/ / / / 4 $\longmapsto$ 4 / / / /

/ / / / / 5 $\longmapsto$ 0 –

Figure 9.7

constructed from strings of tallies with concatenation as the operator; in the integers modulo 5, ///// is equal to the empty string.

Permutations

If we consider the transformations of symmetry of a regular plane geometric figure, these can be expressed as the product of reflections and rotations. For a square, if t represents rotation through 90°, and r represents reflection, we have

$$t^4 = 1$$
$$r^2 = 1$$

The elements of the group can be generated from t and r: $1, t, t^2, t^3, r$ produce from a square ABCD (figure 9.8) the squares ABCD, BCDA,

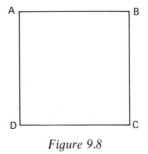

Figure 9.8

CDAB, DABC, DCBA respectively. By inspection we can observe that rt, rt^2, rt^3 respectively produce CBAD, BADC, ADCB. Furthermore rtr produces DABC which is therefore the same as t^3. Thus these permutations are isomorphic to the range of some homomorphism of the group freely generated by the alphabet $\{t, r\}$. The permutations will therefore be isomorphic to the quotient algebra of this homomorphism.

Exercise 9.13

What is rt^2 equivalent to?
Which elements are their own inverses?
How many elements has this group?

Exercise 9.14

(For Rubik's cube addicts.)
   In the Rubik's cube problem, if h is the homomorphism of the free group G generated by $\{R, L, F, B, U, D\}$ to the Rubik's cube group, $R^4, L^4, U^4, B^4, F^4, D^4$ all belong to the kernel. List all the other elements of G which you know belong to the kernel.

Exercise 9.15

If $\mathbb{Z}$ is the group consisting of integers with addition, and $\mathbb{Z}_N$ the group of integers modulo N with addition, what is the kernel of the homomorphism $\mathbb{Z} \rightarrow \mathbb{Z}_N$? Demonstrate that it is a group.

9.9 PROGRAMS AND DATA

Data Representation

We have seen how the data of a problem to be handled by a computer program can be defined in terms of sets, mappings, Cartesian products, etc., and how the operations to be performed on that data can be described in terms of an algebra. For example, the telephone accounts problem described the three data components, *Phone-directory*, and *Owing*, as partial mappings between data types; in the personnel file example the data is represented by sets (*Employee-Id*, *Names*, *Grades* etc.), Cartesian products, mapping (*Employee-id*→(*Names* × *Ages* × *Grades* × *Salaries*)) and so forth; and there are seven operations (CHNAME, GET, ADD, etc.). The transactions on a file problem was similarly treated. In all these problems we would represent their stated structures by programming language constructs. Some examples of these are given in chapter 6 for the telephone accounts problem: an array can be used to represent a mapping whose domain is fixed, or two arrays if the domain may vary over a large set, or an array of records, a list, etc.

The question we wish to ask, having chosen a computer representation of the algebra which is the essence of our problem, is: Is our representation 'adequate'?

Definition 9.11

A representation is *adequate* if (i) every value of our 'abstract' domain— that is, the carriers of our algebra—has corresponding to it at least one representation in our program data which does not also represent any other value in our abstract domain, and (ii) each value in our program data which can be reached by performing the operations and procedures in our program represents some value in our abstract domain.

Example 9.16

A partial finite mapping from Integers to Characters could be represented by an array of records

    array [1:n] **record** I: **Int**

                    B: **Char end**

so that a particular value of this array

    [(37, a),

    (2, c),

    (−3, d),

    (4, z),

    (−6, f)]

would represent a mapping

    $-6 \mapsto f$

    $-3 \mapsto d$

    $2 \mapsto c$

    $4 \mapsto z$

    $37 \mapsto a$

and every finite partial map can be represented by several different array values of this kind: for example

[(2, c),

(−3, d),

(37, a),

(4, z),

(4, z),

(−6, f)]

would represent the same map. The first condition of 'adequacy' is
fulfilled. In order that the second condition is fulfilled we must ensure that
certain kinds of array value cannot be reached by the program's
operations; that is, values such as

[(3, a),

(4, b),

(3, c)]

or

[(3, a),

(4, b),

(4, c)]

which would not clearly represent any specific mapping.

Let us try to formalise this notion with respect to this example. Let A
be the set of values in our abstract domain, and B be the set of values in
our computer data which represents it. B is considered to be a 'model' of
A. We need to define a function $r:B \to A$ such that, given a value b in our
model B, r maps b to the value in A which it represents. Our adequacy
rules can now be stated

(a1) $\forall a \in A . \exists b \in B . r(b) = a$

(b1) $\forall b \in B . \exists a \in A . r(b) = a$

However, these rules are not sufficient: they do not take into account
the fact that not all values of B may be reachable by the operations in the
program. The reachable values may be expressed by means of an
invariant, which is a predicate on the value space. Likewise we may in fact
have invariants which hold over the abstract domain A. Let

Inv-$A : A \to B$

represent the conjunction of all the invariants over A and

Inv-$B : B \to B$

represent the conjunction of all the invariants over B. We can now define

$$A' \triangleq \{x | x \in A \wedge \text{Inv-}A(x)\}$$
$$B' \triangleq \{x | x \in B \wedge \text{Inv-}B(x)\}$$

Now our new adequacy rules become

(a2) $\forall a \in A . \exists b \in B' . r(b) = a$

(b2) $\forall b \in B . \exists a \in A' . r(b) = a$

In our example above, B' represents the set of reachable values of the array, and A is the set of abstract mappings.

Operations

Representation of values is not sufficient. We should require that all the operations in the 'abstract' algebra which captures the essence of our problem are modelled by operations in the computer model of the algebra.

Definition 9.12

Let $Op_A : A_1 \rightarrow A_2$ be an operation in the abstract algebra, where $A_1, A_2 \subseteq A$, and let $Op_B : B_1 \rightarrow B_2$ be the operation in the computer version which is to model Op_A; $B_1, B_2 \subseteq B$.
   Then for the model to have *adequate operations*

$$\forall a \in A_1 . \exists b \in B_1 . r(b) = a \wedge r(Op_B(b)) = Op_A(a)$$

In other words, for every member a of the domain of Op_A there is a member b of Op_B which is matched to it by the 'retrieve' function r such that the results of the operations also match. This is not enough, however. The modelling operation Op_B must not do anything extraneous, that is in addition to what is required by its modelling role: every action carried out by Op_B represents something done by Op_A; that is

$$\forall b \in B_1 . r(Op_B(b)) = Op_A(r(b))$$

This amounts to r being a homomorphism from (B, Op_B) to (A, Op_A). The first statement further stipulates that r is an epimorphism—that is, it is onto A.
   Thus to model our problem algebra correctly in a computer representation, we have to show that there is a retrieve function relating the values in our computer data to those in our problem data which is an epimorphism with respect to the operations defined over each data space.
   The ideas of 'adequacy' and 'retrieve' functions are taken from the work of Jones (1980).

9.10 MORE EXAMPLES

Example 9.17: Permutations and Integers Modulo N

There is an isomorphism between the integers modulo N where N is prime, and the rotations of an $N-1$-sided regular geometric figure. Consider first the rotations of a square again, ABCD (figure 9.9). The rotational permutations are

P1   ABCD

P2   BDCA

P3   CADB

P4   DCBA

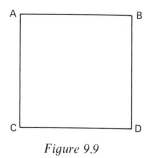

Figure 9.9

Composition of permutations yields the following

$$P2 \circ P3 = ABCD = P1$$
$$P2 \circ P4 = CADB = P3$$
$$P3 \circ P2 = ABCD = P1$$
$$P3 \circ P4 = BDAC = P2$$
$$P4 \circ P2 = CADB = P3$$
$$P4 \circ P3 = BDAC = P2$$
$$P2 \circ P2 = DCBA = P4$$
$$P3 \circ P3 = DCBA = P4$$
$$P4 \circ P4 = ABCD = P1$$

and for all Pn, $P1 \circ Pn = Pn = Pn \circ P1$.

Thus $P = \{P1, P2, P3, P4\}$ is a finite group; P1 is the identity, P2 and P3 are inverses of each other, and P4 is its own inverse. The operation $\circ$ can be defined by the table

o	P1	P2	P3	P4
P1	P1	P2	P3	P4
P2	P2	P4	P1	P3
P3	P3	P1	P4	P2
P4	P4	P3	P2	P1

Compare this with the table defining multiplication of non-zero integers modulo 5

*	1	2	3	4
1	1	2	3	4
2	2	4	2	3
3	3	2	4	2
4	4	3	2	1

Inspection shows that there is a clear isomorphism matching n with P_n, such that

$$\forall k, n, m \in \{1, 2, 3, 4\} . P_k = P_n . P_m \Leftrightarrow k = n*m$$

In other words, all the cells of the two tables match.

The fact that there is a homomorphism from the non-zero integers with multiplication to the integers modulo N with multiplication means that, using the rules of composition of homomorphisms, there is a homomorphism from non-zero integers with multiplication to rotational permutations of $N-1$ points (where N is prime).

This can be illustrated with the diagram in figure 9.10. Figure 9.10 also

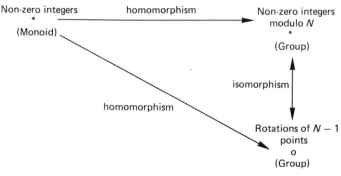

Figure 9.10

illustrates the fact that if *h* is a homomorphism from algebra *A* to algebra *B*

$$h: A \rightarrow B$$

all the rules of *A* also hold in *B*, but therefore *B* may have extra rules in addition to those reflected by *h* from *A*. Thus, we can have a homomorphism from a Monoid to a Group but not from a Group to a Monoid which is not a group. Groups have extra 'rules', namely that all elements have inverses.

This may seem a little odd at first, for one normally regards a group as in some way more 'sophisticated' than a monoid; there are more theorems to be discovered with groups than monoids. Yet the codomain of a homomorphism is generally smaller than the domain. However, one may deduce that there are therefore more assertions to be made about the codomain; certainly at least as many.

Example 9.18: Logarithms

Logarithms illustrate a useful isomorphism between the Reals > 0 with *
and the Reals with +. Consider figure 9.11. Logarithms to the base *b* give

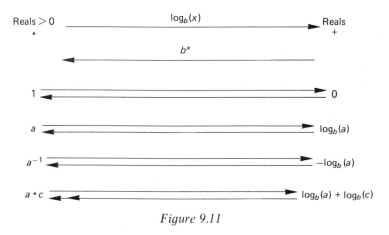

Figure 9.11

a group isomorphism: the identity element (1) is paired with the identity element (0), inverses (a^{-1}) with inverses ($-a$) and the homomorphism rule holds in each direction

$$\log_b(a*c) = \log_b(a) + \log_b(c)$$
$$b^a * b^c = b^{(a*c)}$$

Example 9.19: Car Supply Database

Recalling our car database, there were a number of facets defining the car

  B body styles

  C colours

  Eng engine sizes

  T trim styles

and the set of possible cars was

  $B \times C \times Eng \times T$

A stock control database could record the numbers of each car in stock by data of the type

  $Db = (B \times C \times Eng \times T) \to \mathbb{N}_0$

Then a few operators could be defined

  $Init: \to Db$

  Add: $B \times C \times Eng \times T \times \mathbb{N} \times Db \to Db$

  How many: $B \times C \times Eng \times T \times Db \to \mathbb{N}_0$

The axioms or rules of this system would be

1. How many $(b, c, e, t, \text{Init}(\ )) = 0$
2. How many $(b, c, e, t, \text{Add}(b, c, e, t, n, d)) = $ How many $(b, c, e, t, d) + n$

This is rather an oversimplified system because there is no way of withdrawing stock from it! However, let us work with it for the time being.

 Homomorphisms can be a way of forgetting or ignoring details. Suppose we are not interested in interior trim of the cars in stock. We can construct another system, the homomorphic image of the one above, as follows

  $Db_T = (B \times C \times Eng) \to \mathbb{N}_0$

$Init_T$, Add_T, How many$_T$ have schemes as before with T removed. The rules of the system are similar, but with t omitted

1. How many$_T$ $(b, c, e, \text{Init}_T(\ )) = 0$
2. How many$_T$ $(b, c, e, \text{Add}_T(b, c, e, n, d)) = $ How many$_T$ $(b, c, e, d) + n$

 A homomorphism $h_T: Db \to Db_T$ is such that

$$h_T(db) = [\langle b, c, e \rangle \mapsto \sum_{t \in T} db(\langle b, c, e, t \rangle | b \in B \wedge c \in C \wedge e \in Eng]$$

(recall that the database is of type $B \times C \times Eng \times T \to \mathbb{N}_0$.)

Exercise 9.16

To show that h_T above is a genuine homomorphism, what has to be proved? Show (informally) that h_T is a homomorphism.

Other 'forgetful' homomorphisms, h_B, h_C, h_{Eng} and combinations of them can be defined.

Each homomorphism defines a quotient algebra, whose carrier is a set of equivalence classes.

Exercise 9.17

Describe, in words, what are the equivalence classes of the 'cars', $\langle b, c, e, t \rangle$, associated with the quotient algebra generated by h_T.

Exercise 9.18 (Discussion/Workshop)

This exercise is suitable for group or class discussion, or for a workshop exercise.

In the personnel file examples, last encountered towards the end of chapter 8, we showed how the system could be regarded as an algebra, with carriers

{*Employees, Employee-Id, Names, Ages, Grades, Salaries*}

and operations

DELETE, ADD, GET, CHNAME, CHAGE, CHSALARY, CHGRADE

One can imagine that certain users of the system may be allowed to do some operations and not others: for example, Personnel may DELETE, ADD, GET, CHNAME, CHAGE and Accounts may GET, CHSALARY, CHGRADE perhaps.

Let us assume three systems: a universal one containing all the information as described in chapter 7, a system for Accounts which has the operations GETA, CHSALARY, CHGRADE and a system for Personnel which has operations DELETE, ADD, GETP, CHNAME, CHAGE. Further we shall assume that while personnel may define a person's initial Salary and Grade, these are not their concern and so the GETP operation returns Names × Ages. Likewise the Ages is no concern of Accounts and so GETA returns a result of type Names × Grades × Salaries.

The Personnel subsystem and Accounts subsystem are homomorphic images of the universal system. Discuss the nature of these two homomorphisms h_P, h_A.

Think of likely axiomatic rules relating ADD and GET, CHNAME and GET, CHAGE and GET, CHGRADE and GET, CHSALARY and GET in the universal system, and appropriate rules in the subsystems relating ADD and GETP, CHGRADE and GETA etc. Show that the homomorphisms h_P and h_A preserve the rules of the subsystems which form their codomains.

Exercise 9.19 (Workshop)

At the end of chapter 8, exercises relating to each of the four workshop projects (Nuclear power-station, Personal diary, Space-borne computer system, Fault-tolerant control system) were given. In each one you were asked to define operations and their schemes so as to define an algebra. Whichever workshop problem you are engaged in, first make sure that a proper complete signature has been defined (chapter 9 has illustrated some more signatures which will be a guide). Homomorphisms can now be used in two ways: firstly to define good computer representations of the problem data, and secondly to define subsystems of the total system. Therefore

(1) Assuming you only have the computer programming constructs of Arrays (maps from a finite subsequence of integers or an enumerated set to another data type), Records (fixed finite Cartesian products), Integers, Booleans, Reals, and Enumerated sets (fixed finite sets of names), define a homomorphic algebra to that of your problem which uses these computer constructs as types. The operations must have schemes taken from the programming language data types described and should therefore reflect the specification of procedures (which may return results) performing the desired operation. Define the homomorphism from your model to the abstract algebra, consider any properties of the algebra it should preserve (invariants over the data, rules about the operations) and show what steps must be taken to ensure these properties are preserved.

(2) Define one or two more subsystems and the associated 'forgetful' homomorphisms. For example in the diary problem, read-only operations could be made available as a subsystem, or some user may be allowed to add items but not read, delete or move them. In the nuclear power station, reading and controlling operations likewise can be separated. In the space-borne computer problem, navigation and control functions can be separated.

SUMMARY OF RESULTS

Result	Reference
Homomorphisms map identities to identities	Theorem 9.1
Homomorphisms map inverses to inverses	Theorem 9.2
Homomorphisms preserve associativity	Theorem 9.3
$h(|A|)$ is closed under the operators of A	Theorem 9.4
The equivalence classes generated by a homomorphism are in one–one correspondence with the members of its range	Theorem 9.5
The kernel of a group homomorphism is a sub-group	Theorem 9.6
If $h:F \to G$ and $a \in F$ and p a member of the kernel of h, then $h(a) = h(p+a) = h(a+p)$	Theorem 9.7
Converse of theorem 9.7	Theorem 9.8
If h, k are homomorphisms, then so is $h \circ k$	Theorem 9.9

Chapter 10 Functions and their Specifications

NOTATION INTRODUCED

Notation	Concept	Reference
pre, post	pre-condition, post-condition, implicit specification	Def. 10.1
let...in, where	local definitions	Section 10.2
	consistent, inconsistent specifications	Def. 10.2
	complete specifications	Def. 10.3

In chapter 6 we learned about functions and mappings as data objects. Functions with infinite domains, although in principle the same kinds of objects as ones with finite domains, are at least intuitively different, and in fact also have some important theoretical differences.

In chapter 6 we largely restricted our attention to functions with finite domains, since these can be more easily represented by finite pieces of computer data. In this chapter we shall consider functions with potentially infinite domains. These are very much akin to, indeed scarcely distinguishable from, operations in an algebra.

Consider, for example, the function 'to the power of' on the real numbers. This has a type or scheme

$$\mathbb{R} \times \mathbb{R} \to \mathbb{R}$$

and we could try to define it using the kind of explicit definition we considered in chapter 6. Such an explicit definition would be in terms of already defined operations and functions. An operator or function defined in such a way is often called 'derived' for this reason.

However, we may well find difficulty in producing an explicit definition for a number of reasons: the result of the function may not be expressible in a single formula, but may require a number of assertions to define its value appropriately; we may even not be interested in the exact value of the function but may be satisfied with an approximation. For such reasons we introduce (implicit) specifications of functions.

214

10.1 IMPLICIT SPECIFICATIONS

Definition 10.1

The *pre-condition* of a function specification limits the domain values by making assertions about them. Instead of defining exactly what the value of a function should be for given input arguments, an assertion may be made about the relationship between the domain values and the range values, which similarly constrains the range values for given arguments. Such an assertion is called a *post-condition*. A function which is specified by means of a pre-condition and a post-condition is said to have an *implicit specification*.

Example 10.1

Consider a square-root function. In general, $\sqrt{x}$ is not exactly representable with a finite amount of information, and so an approximation has to be provided by any mechanical algorithm. Provided the result is within a required tolerance, however, one need not be interested in the exact function. The type clause is

$$Sqrt: \mathbb{R} \times \mathbb{R} \to \mathbb{R}$$

Here the first argument is the number whose square root is sought, and the second argument is the tolerance. The tolerance must be greater than zero and we cannot find a square root of a negative number, so the pre-condition is

$$\textbf{pre-}sqrt\,(x, t) \triangleq x \geqslant 0 \wedge t > 0$$

Observe once again that the only variables occurring on the right-hand side of the definition symbol $\triangleq$ are x and t which are bound by the parameters on the left-hand side.

The post-condition now states that the result must be within tolerance. The post-condition is a function of the input arguments and the results

$$\textbf{post-}sqrt: \mathbb{R} \times \mathbb{R} \times \mathbb{R} \to \mathbb{B}$$
$$\textbf{post-}sqrt\,(x, t, r) = r > 0 \wedge x - t < r^2 < x + t$$

If we had not asserted in the post-condition that $r \geqslant 0$ the implementor of the function would have been free to choose a negative approximation to the square root. This may or may not be acceptable.

Theorem 10.1

All explicit definitions can be rephrased as implicit specifications.

Proof (*sketch*)

Let the definition of the function be of the following form

$$f(x_1, \ldots, x_n) = E$$

where E is an expression in which x_i, $1 \leqslant i \leqslant n$ occur free. This can be replaced by a post-condition as follows

post-$f(x_1, \ldots, x_n, r) \triangleq r = E$

where r is a symbol not the same as any of the x_i.   □

Example 10.2

In the dels function, we can rephrase the specification implicitly as follows

dels: $A^* \times \mathbb{N} \rightarrow A^*$

pre-dels $(l, n) \triangleq n \leqslant$ len l

post-dels $(l, n, r) \triangleq r =$ **if** $n = 1$ **then** tl l
$\qquad\qquad\qquad\qquad$ **else** $\langle$ hd $l \rangle \| $ dels $(\text{tl } l, n - 1)$

This post-condition may be expressed a little more naturally as follows

post-dels $(l, n, r) \triangleq$ **if** $n = 1$ **then** $r =$ tl l
$\qquad\qquad\qquad$ **else** $r = \langle$ hd $l \rangle \|$ dels $(\text{tl } l, n - 1)$

Again, since the type of the post-condition is determined by that of the function, it is usually omitted.

Exercise 10.1

Replace the definition of mods, front and subs as implicit specifications using post-conditions.

Example 10.3

In sorting and other applications, we often wish to express the fact that one sequence is a permutation of another. Let us define a function called 'is-permutation'. Its type will be

is-permutation: $A^* \times A^* \rightarrow \mathbb{B}$

Such a function, whose result is of type Boolean, is called a 'predicate'. As a convention we shall often call such functions by names starting with 'is-'. For the lists to be permutations of each other, their lengths must be the same, each element of the first list must be found in the second, and the cardinality of its occurrences must be the same.

The first condition is simply

$$\text{len } l1 = \text{len } l2$$

The second can be stated

$$\forall i \in \{1 \ldots \text{len } l1\} . \exists j \in \{1 \ldots \text{len } l2\} . l1(i) = l2(j)$$

The third condition is more complex

$$\forall i \in \{1 \ldots \text{len } l1\} . \text{card } \{j \mid j \in \{1 \ldots \text{len } l1\} \wedge l1(j) = l1(i)\}$$
$$= \text{card } \{j \mid j \in \{1 \ldots \text{len } l2\} \wedge l2(j) = l1(i)\}$$

Hence the whole specification is

$$\text{is-permutation:} A^* \times A^* \to \mathbb{B}$$
$$\text{is-permutation } (l1, l2) \triangleq \text{len } l1 = \text{len } l2 \wedge$$
$$(\forall i \in \{1 \ldots \text{len } l1\} . \exists j \in \{1 \ldots \text{len } l2\} . l1(i) = l2(j)) \wedge$$
$$(\forall i \in \{1 \ldots \text{len } l1\}$$
$$\text{card } \{j \mid j \in \{1 \ldots \text{len } l1\} \wedge l1(j) = l1(i)\}$$
$$= \text{card } \{j \mid j \{1 \ldots \text{len } l2\} l2(j) = l1(i)\})$$

We can now define a data type, which is the set of all permutation functions on A^*. Any member of this type will be a function which takes a list of A elements and produces another list of A elements which is a permutation of the one given

$$\text{Per} = \{ p \mid p \in (A^* \to A^*) \wedge (\exists n \in \mathbb{N}_0 . \text{dom } p = A^n$$
$$\wedge (\forall a \in A^n . \text{is-permutation } (a, p(a))))\}$$

Certain permutations are their own inverses; that is, when applied twice to a list, they leave the list unchanged, or produce the identity permutation. (Recall from chapter 9 that the set of all permutations of a fixed number n of objects forms a group with an identity and inverses with respect to composition of permutations.) We can define a predicate

$$\text{is-own-inverse:} \text{Perm} \to \mathbb{B}$$

post-is-own-inverse $(p, b) \triangleq$
$$b \Leftrightarrow (\forall a \in \text{dom } p . p(p(a)) = a)$$

Examples of permutations which are their own inverses are those consisting of disjoint transpositions

$$\langle A, B, C, D \rangle \mapsto \langle B, A, C, D \rangle$$
$$\langle A, B, C, D \rangle \mapsto \langle C, D, A, B \rangle$$

and the identity permutation itself

$$\langle A, B, C, D \rangle \mapsto \langle A, B, C, D \rangle$$

and others.

10.2 LOCAL DEFINITIONS

It is sometimes convenient to define either a value or a function with local scope. Whenever we use an expression, it will be in the scope of a number of names of variables, functions, etc.

Example 10.4

Consider the post-condition for dels which we last formulated

> **post**-dels $(l, n, r) \triangleq$ **if** $n = 1$ **then** $r = $ tl l
>
> **else** $r = \langle$hd $l\rangle \| $dels (tl $l, n - 1)$

The expression to the right of the $\triangleq$ definition symbol is within the scope of the names l, n, r. If we wished, we could define a new symbol standing for the value tl l. This would be permissible because the scope of l is the expression following $\triangleq$, and we can define a new symbol either by means of a **let** clause preceding the scope or a **where** clause following it

> **post**-dels $(l, n, r) \triangleq$ **let** $t \triangleq$ tl l **in**
>
> **if** $n = 1$ **then** $r = t$ **else** $r = \langle$hd $l\rangle \| $dels$(t, n - 1)$

The scope of t now is the expression following 'in'.
  The above may alternatively be expressed

> **post**-dels $(l, n, r) =$
>
> **if** $n = 1$ **then** $r = t$ **else** $r = \langle$hd $l\rangle \| $dels$(t, n - 1)$
>
> **where** $t \triangleq$ tl l

In the former case t could not be referenced in this context before the let clause, or in the latter case after the where clause.
  Clearly this becomes more valuable if tl l is a much longer expression.

Example 10.5

To illustrate local definition of functions, let us look at the definition of is-permutation. This can be simplified if we define two functions

> inds: $A^* \to \mathscr{P}\mathbb{N}$
>
> inds $(l) \triangleq \{1 \ldots$ len $l\}$

and

multp: $A^* \to \mathbb{N}_0$

multp $(l) =$ card $\{j \mid j \in \text{inds}\,(l) \wedge l(j) = l1(i)\}$

In the second case the definition of multp contains references to $l1$ and i. It therefore has meaning only if it is defined within the scope of $l1$ and i. It must also be defined within the scope of the definition of inds.

Our definition of is-permutation can now be expressed as follows

is-permutation: $A^* \times A^* \to \mathbb{B}$

is-permutation $(l1, l2) \triangleq \text{len } l1 = \text{len } l2 \wedge$

    let inds: $A^* \to \mathscr{P}\mathbb{N}$

        inds $(l) \triangleq \{1 \ldots \text{len } l\}$ **in**

    $(\forall i \in \text{inds}\,(l1).\ \exists j \in \text{inds}\,(l2).\,l1(i) = l2(j)) \wedge$

    $(\forall i \in \text{inds}\,(l1).$

    let multp: $A^* \to \mathbb{N}_0$

        multp $(l) \triangleq$ card $\{j \mid j \in \text{inds}\,(l) \wedge l(j) = l1(i)\}$

    in multp $(l1) =$ multp $(l2))$

In the definition of multp, i is bound to the second universal quantification $(\forall i \in \text{inds}(l1))$, $l1$ is bound to the argument $l1$ on the left of the $\triangleq$ defining is-permutation, and inds is within the scope and bound by the **let** inds... clause.

10.3 EXAMPLES

We have enough notational apparatus to apply these ideas to some of our more substantial thematic examples.

Example 10.6: Transactions on a File

In this example the sorts for the operations were

    $T = \{File, R, INF\}$

and the operations were

    Purge: $File \to File$

    Delete, Add: $File \times R \to File$

    Update: $File \times R \times INF \to File$

    Inspect: $File \times R \to INF \cup \{\text{Error}\}$

The operations may be considered as functions with domains and codomains as indicated. However, to define these in a functional style, we need to define the types. A possible definition for a model of these types (see chapter 9) is

$$File_c = (R_c \times INF_c)^*$$

The other types can be left as arbitrary disjoint sets.

We can then define the functions as

$$Purge_c : File_c \rightarrow File_c$$

$$Purge_c(f) = \langle \ \rangle$$

Thus purge empties the file

$$Delete_c : File_c \times R_c \rightarrow File_c$$

pre-$Delete_c(f, r) \triangleq (\exists i \in \{1 \ldots \text{len } f\} . \exists I \in INF . f(i) = \langle r, I \rangle)$

post-$Delete_c(f1, r, f2) \triangleq \exists i \in \{1 \ldots \text{len } f1\} . \exists I \in INF .$

$$f1(i) = \langle r, I \rangle \wedge f2 = \text{dels}(f1, i)$$

Here $Delete_c$, by its pre-condition, is only defined for a record r which is already present in the file. The post-condition states that the new file consists of the file with the indicated record deleted.

$$Add_c : File_c \times R_c \rightarrow File_c$$

pre-$Add_c(f, r) \triangleq \sim (\exists i \in \{1 \ldots \text{len } f\} . \exists I \in INF . f(i) = \langle r, I \rangle)$

post-$Add_c(f1, r, f2) \triangleq f2 = f1 \| \langle r, INITI \rangle$

$INITI \in INF_c$ is an 'initial value' of the type INF_c. The pre-condition states that r is not already present in the file and the post-condition states that the new file consists of the old file with $\langle r, INITI \rangle$ appended to it.

$$Update_c : File_c \times R_c \times INF_c \rightarrow File_c$$

pre-$Update_c(f, r, I) \triangleq (\exists i \in \{1 \ldots \text{len } f\} . \exists J \in INF_c . f(i) = \langle r, J \rangle)$

post-$Update_c(f1, r, f2) \triangleq \text{len } f2 = \text{len } f1 \wedge$

$$(\exists i \in \{1 \ldots \text{len } f1\} . (\exists J \in INF_c . f1(i) = \langle r, J \rangle \wedge f2(i) = \langle r, I \rangle) \wedge$$

$$(\forall j \in \{1 \ldots \text{len } f1\} - \{i\} . f2(j) = f1(j)))$$

The pre-condition states that r is present in the file. The post-condition states that the new file is the same as the old, with the exception of the record containing r, which has I as its information field instead of the value found previously.

$$Inspect_c : File_c \times R_c \rightarrow INF_c \cup \{\text{Error}\}$$

pre-$Inspect_c(f, r) \triangleq \text{True}$

post-$Inspect_c(f, r, k) \triangleq$

$$(\exists i \in \{1 \ldots \text{len } f\} . \exists I \in INF_c . f(i) = \langle r, I \rangle \Rightarrow k = I) \wedge$$

$\sim (\exists i \in \{1 \ldots \text{len } f\} . \exists I \in INF_c . f(i) = \langle r, I \rangle)$

$\Rightarrow k = \text{Error}$

Retrieve functions on the types and operations of this model should preserve the rules of the abstract algebra, such as

Inspect (*Update* $(Add(f,r),r,i),r) = i$

Example 10.7: Car Supply Database

This example relates to our stock of cars, with

B	body styles
C	colours
Eng	engine sizes
T	trim styles

The sorts were $\{Db, B, C, Eng, T, \mathbb{N}, \mathbb{N}_0\}$ where

$Db = (B \times C \times Eng \times T) \to \mathbb{N}_0$

The operations are

$\{Init, Add, How\ many, Withdraw\}$

and their schemes and specifications, regarded as functions, are (we will omit the subscript c to indicate the model algebra for notational simplicity)

Init: $\to Db$

pre-*Init*( ) = True

post-*Init* $(d) \triangleq d = [x \mapsto 0 | x \in (B \times C \times Eng \times T)]$

The operation Init initialises the database so that all combinations of bodies, colours etc. are mapped to 0.

Add: $B \times C \times Eng \times T \times N \times Db \to Db$

pre-*Add* $(b, c, e, t, n, d) \triangleq$ True

post-*Add* $(b, c, e, t, n, d1, d2) \triangleq$

$\qquad (\forall x \in (B \times C \times Eng \times T) . x \neq \langle b, c, e, t \rangle \Rightarrow d2(x) = d1(x)) \wedge$

$\qquad d2(\langle b, c, e, t \rangle) = d1(\langle b, c, e, t \rangle) + n$

How many: $B \times C \times Eng \times T \times Db \to \mathbb{N}_0$

pre-*How many* $(b, c, e, t, d) \triangleq$ True

post-*How many* $(b, c, e, t, d, n) \triangleq d(\langle b, c, e, t \rangle) = n$

Withdraw: $B \times C \times Eng \times T \times Db \times \mathbb{N} \to Db$

pre-$Withdraw\,(b,c,e,t,d,n) \triangleq n \leqslant d(\langle b,c,e,t \rangle)$

post-$Withdraw\,(b,c,e,t,d1,n,d2) \triangleq$

$$d1(\langle b,c,e,t \rangle) = n + d2(\langle b,c,e,t \rangle)$$

This withdraws n vehicles of type $\langle b,c,e,t \rangle$, provided that the database contains at least n such.

Exercise 10.2

Informally demonstrate that the above model satisfies the axioms given in chapter 9, namely

(i)   *How many* $(b,c,e,t,Init(\;)) = 0$

(ii)   *How many* $(b,c,e,t,Add(b,c,e,t,n,d)) =$

How many $(b,c,e,t,d) + n$

and an additional axiom for Withdraw.

(iii) *How many* $(b,c,e,t,d) \geqslant n \Rightarrow$

How many $(b,c,e,t,Withdraw(b,c,e,t,d,n)) =$

How many $(b,c,e,t,d) - n$

Example 10.8: Personnel File

In this example the types were

{*Employees, Employee-Id, Names, Ages, Grades, Salaries*}

We shall model *Employees* as

Employees = Employee-Id $\mapsto$ *Names* $\times$ *Ages* $\times$ *Grades* $\times$ *Salaries*

Where $\mapsto$ indicates a partial mapping. The operations can now be specified as follows

$DELETE: Employees \times Employee\text{-}id \rightarrow Employees$

pre-$DELETE\,(E,eid) \triangleq eid \in \mathrm{dom}\,E$

post-$DELETE\,(E1,eid,E2) \triangleq$

$(\forall e \in \mathrm{dom}\,E1\,.\,e \neq eid \Rightarrow E2(e) = E1(e)) \wedge$

$\mathrm{dom}\,E2 = \mathrm{dom}\,E1 - \{eid\}$

eid is removed from the domain of the employees.

$ADD: Employees \times Employee\text{-}id \times Names \times Ages \times$

$Grades \times Salaries \rightarrow Employees$

pre-$Add\,(E,eid,n,a,g,s) \triangleq eid \notin \mathrm{dom}\,E$

post-$Add\,(E1, eid, n, a, g, s, E2) \triangleq$
$$\qquad \text{dom } E2 = \text{dom } E1 \cup \{eid\} \;\wedge$$
$$\qquad (\forall e \in \text{dom } E1 \,.\, E2(e) = E1(e)) \;\wedge$$
$$\qquad E2(eid) = \langle n, a, g, s \rangle$$

The effect is to add $eid \mapsto \langle n, a, g, s \rangle$ to employees E.

Exercise 10.3

Complete the specifications of

$\qquad GET{:}\, Employees \times Employee\text{-}id \rightarrow Names \times Ages \times Grades \times Salaries$

$\qquad CHNAME{:}\, Employees \times Employee\text{-}id \times Names \rightarrow Employees$

$\qquad CHAGE{:}\, Employees \times Employee\text{-}id \times Ages \rightarrow Employees$

$\qquad CHGRADE{:}\, Employees \times Employee\text{-}id \times Grades \rightarrow Employees$

$\qquad CHSALARY{:}\, Employees \times Employee\text{-}id \times Salaries \rightarrow Employees$

Exercise 10.4

Show that the model of the personnel file produced here satisfies the axiomatic rules for the system devised in exercise 9.18.

On Axioms and Models

In chapters 8 and 9 we saw how to define an algebra in terms of its sorts and operator schemes, and how the effects of the operators can be related to each other by means of axiomatic rules. In this chapter we have seen how functions, and hence operators in an algebra, which are essentially the same, can be specified by defining their pre-conditions and post-conditions in terms of a model based on sets, lists, mappings etc.

The relationship between these two styles is that the axiomatic definitions are the more abstract versions: each model should satisfy the axioms if it is a true model, and in fact there is a variety of different possible models. In the last analysis everything can be modelled with binary strings, but the operations become tedious and complex.

For this reason, present-day research into specification techniques tends to favour an axiomatic style, since there is less prejudice on what model should be taken for the eventual computer representation. The work of Burstall and Goguen (1980a, 1980b, *q.v.* for further references) investigates the consequences of adhering to this style of specification, and is essential reading for anyone who wishes to pursue this aspect of the subject to any depth.

10.4 SUMMARY

Algebras, Specifications and Programs (Discussion)

Informally, specifications may be considered as a class of functions which have

$$Sys = \{\ldots\} \quad \text{—All systems}$$

as domain and

$$B = \{\text{Satisfies, Does not satisfy}\}$$

as ranges. Thus the class of all specifications is

$$Spec = Sys \rightarrow B$$

To make this a little less informal we can say that Sys is the collection of algebras, that is any member of Sys is a collection of sets with functions or operations defined on them. All the examples given in section 10.3 are algebras.

These algebras or systems are candidates for satisfying some specification. A specification is thus a list of constraints, conditions or 'axioms' which the system must satisfy.

First, the signature, comprising the sort names, operator names and operator schemes must be the same. Then the operators of the algebra must satisfy the axioms or rules given in the specification.

In section 10.3 and in chapter 9 we referred to these algebras as models because they are model algebras which satisfy the specifications.

A computer representation of one of these algebras will be a direct representation with programming language functions and procedures representing the operations of the algebra, and variables, arrays and other types of data representing the sorts.

There will be homomorphisms between some of these various algebras, all of which satisfy a given specification, as we saw in chapter 9. Many of these algebras will be isomorphic to each other. The diagram in figure 10.1 illustrates this.

Consistency and Completeness

Definition 10.2
A specification could impose such restrictive axioms that no algebra, except perhaps for one whose carriers include the empty set, can satisfy it. Such a specification is called 'inconsistent'; otherwise it is called 'consistent'.

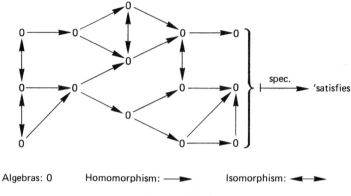

Algebras: 0   Homomorphism: ——▶   Isomorphism: ◀——▶

Figure 10.1

Definition 10.3
A specification is termed 'complete' if it is restrictive enough that only one algebra (or system) satisfies it, to within isomorphism. That is, it is complete if all algebras satisfying it are isomorphic (hence essentially the same as each other).

We always seek to formulate consistent specifications, and we often wish to formulate consistent and complete specifications.

Example 10.9

Because systems are modelled by algebras, we can specify abstract algebraic systems in just the same way as we specify information systems. For example, Boolean algebras are any algebras which satisfy the following specification

$$\langle T, \Omega \rangle$$

where

$$T = \{C\}$$
$$\Omega = \{\cap : C \times C \to C$$
$$' : C \to C\}$$

Axioms

$(\forall x, y \in C . x \cap y = y \cap x)$

$(\forall x, y, z \in C . x \cap (y \cap z) = (x \cap y) \cap z)$

$(\forall x, y, z \in C . \textbf{if } x \cap y' = z \cap z' \textbf{ then } x \cap y = x)$

$(\forall x, y, z \in C . \textbf{if } x \cap y = x \textbf{ then } x \cap y' = z \cap z')$

This specification is satisfied by many algebras, including the familiar

Boolean algebra with two truth values, True and False. It is therefore consistent but not complete. We can make it complete by also stipulating that, for example

Card $C = 2$

or

Card $C = 4$

But if we required:

Card $C = 5$

we would have a specification which is inconsistent.

Gödel's theorem, formulated by Gödel in 1931 (Gödel, 1931), of which a highly readable account is given by Nagel and Newman (1959), states, in essence, that no specification for the integers can be both consistent and complete.

Exercise 10.5 (Workshop)

Write specifications for the operation and functions of your system. Thus, relating back to chapter 8, exercises 8.12, 8.13, 8.14 and 8.15

(1) For the space-borne computer system, specify the operation which calculates the future position, future velocity and other operations which were defined in exercise 8.12.
(2) For the fault-tolerant computer system, specify the functions for the checkers and arbiters and for the whole system, whose signatures and schemes were produced for exercise 8.13.
(3) For the diary problem, specify the operations for adding, deleting and moving an event, and for any other operation whose scheme was defined in exercise 8.14.
(4) For the nuclear power station, specify the algebraic operators which were identified and whose schemes were defined in exercise 8.15.

Exercise 10.6 (Workshop)

In exercise 9.19 part (1), a model algebra homomorphic to that produced in the exercises at the end of chapter 8 was produced, which uses computer constructs as types. Define specifications for the operations in this model algebra, and attempt to construct arguments to demonstrate that they fulfil the requirement that one algebra is a model of the other— that is, the two rules given in definitions 9.11 and 9.12.

SUMMARY OF RESULTS

Result	Reference
All explicit function definitions can be rephrased as implicit specifications	Theorem 10.1

PART IV: Specification and Correctness of Programs

In a sense this final part is the culmination of the use to which discrete mathematics can be put in software engineering. Discrete mathematics permits us to state precisely and in abstract terms what we mean. The occasion in the software engineering process where this activity becomes most clearly useful is in specification. Specification is the statement of what the putative, intended product is to do: its function, its attributes, a statement of what it does and its required performance without prescribing how it is to do it, its internal structure or algorithm; in other words, a description of its behaviour and that alone.

Discrete mathematics is a very appropriate linguistic tool for expressing such specifications, for we can define function specifications, data types and operations without prescribing algorithms. Furthermore the language of logic, a linguistic keystone studied early on in chapter 3 enables us to ascertain, by providing a proof, that a design fulfils a specification. Progression from one stage of the software engineering process to the next, confident of the correctness of the progression, depends on a mechanism for proving the correctness of the transformations being available.

Chapter 12 describes a way of defining specifications of programs, using pre-conditions and post-conditions, and summarises proof techniques based on the semantics of programming languages. Preliminary to this however, we do a little more work on relations, for one aspect of proofs in the context of pre-conditions and post-conditions is the fact that statements or assertions about programs have a special relation with each other and form a structure known as a lattice. It is necessary to demonstrate this structure in order to put into context an important concept used in proving that programs meet their specifications, namely that of 'weakest pre-conditions'.

Part IV therefore starts with an extension to the concepts of relations and finishes with the final chapter on specification and correctness.

Chapter 11 Relations and Lattices

NOTATION INTRODUCED

Notation	Concept	Reference
aRb	binary relation	Def. 11.1
$\text{Graph}(R)$	graph of a relation	Def. 11.2
	reflexive, symmetric, transitive	Def. 11.3
	equivalence relation	Def. 11.4
	closure	Defs 11.5, 11.6
	transitive closure	Def. 11.7
	antisymmetric	Def. 11.8
	asymmetric	Def. 11.9
	partial order, poset	Def. 11.10
	diagrams representing partial orders	Section 11.3
	linear orders	Def. 11.11
R^*, R^+	reflexive transitive closure, transitive closure	Section 11.3
$aRS, a \geqslant S$	lower bound	Def. 11.12
$SRb, S \leqslant b$	upper bound	Def. 11.13
glb, $\sqcap$	greatest lower bound	Def. 11.14
lub, $\sqcup$	least upper bound	Def. 11.15
	lattice	Def. 11.16
	complete lattice	Def. 11.17
$\sqsubseteq$	lattice relation	Section 11.4
$\top \perp$	top, bottom	Section 11.4
	monotonic	Def. 11.18
	continuous	Def. 11.19

11.1 RECAPITULATION

Let us recall from chapter 7 the definition of relations and the associated notation.

Definition 11.1
A binary relation R is a predicate of two arguments

$$R: A \times B \to \mathbb{B}$$

Definition 11.2
The graph Graph(R) of a relation R is the set of pairs

$$(a, b) \in A \times B$$

such that

$$(a, b) \in \text{Graph}(R) \Leftrightarrow R(a, b) = \text{True}$$

Notation

We write

$$aRb$$

if

$$R(a, b) = \text{True}$$

Given a relation on a set A, that is

$$R = A \times A \to \mathbb{B}$$

various properties were defined.

Definition 11.3
The relation R is *reflexive* if

$$\forall a \in A . aRa$$

The relation R is *symmetric* if

$$\forall a, b \in A . aRb \Leftrightarrow bRa$$

The relation R is *transitive* if

$$\forall a, b, c \in A . aRb \wedge bRc \Rightarrow aRc$$

Finally in section 7.3 we defined equivalence relations.

Definition 11.4
A relation

$$R = a \times a \to \mathbb{B}$$

is an *equivalence relation* if it is reflexive, transitive and symmetric.

Examples of equivalence arise out of quotient algebras which we explored in chapter 9.4.

Example 11.1

Recall (section 9.4) that for any homomorphism into an algebra B, that is

$$h: A \to B$$

then there results a quotient algebra A/h. One carrier of this algebra is the partition of A resulting from h, and in fact it is the set of equivalence classes corresponding to the equivalence relation generated by h, namely

$$E_h \subseteq A \times A$$
$$a E_h b \Leftrightarrow h(a) = h(b)$$

Example 11.2

In chapter 9, examples 9.19 were given to show how homomorphisms can be used to 'forget' details of a set of attributes. In all such cases equivalence classes and relations result. In fact, given any mapping or homomorphism $h: A \to B$, equivalence classes and relations are generated as above

$$E_h \subseteq A \times A$$
$$a E_h b \Leftrightarrow h(a) = h(b)$$

regardless of whether the domain A and range B are familiar algebras. Hence in the database system for cars which was considered in example 9.19, a homomorphism h was defined which 'forgot' the interior trim, and the equivalence relation resulting would be

$$\langle b, c, e, t_1 \rangle E_h \langle b, c, e, t_2 \rangle$$

Example 11.3

A product number suitable for computer programs has the format shown in figure 11.1 Equivalence relations abound: aRb if a and b have the same Product type/Product number/Variant/Medium/Format/Edition or any combination of these. Again a 'forgetful' homomorphism is involved, typically forgetting say the format and medium.

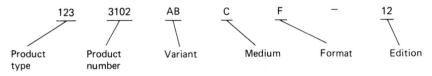

Figure 11.1

Example 11.4

We wish to match strings according to the following criteria: leading and trailing spaces and new lines are to be ignored; new lines are equivalent to spaces; multiple spaces and/or new lines are equivalent to a single space; upper case letters are equivalent to their lower case counterparts.

A program which implemented such a matching process would be an implementation of an equivalence relation; given a pair of strings it would respond with True if they matched and False otherwise. The fact that the relation is an equivalence relation would result in the following behavioural characteristics of the program SM(a, b)

SM(a, a)   gives True

SM(a, b)   gives the same result as SM(b, a)

If SM(a, b) gives True and SM(b, c) gives True then

SM(a, c) gives True

11.2 CLOSURES OF RELATIONS

Definition 11.5
If we have a relation R, we can define a new relation R' which is the 'closure' of R with respect to some particular property of relations. Thus, symmetry being a property of relations, given R we can define the 'symmetric closure' of R, which is the smallest relation R_s which is symmetric and which includes R.

The terms 'smallest' and 'includes' may be defined in terms of the graphs of relations. Recall that graphs of relations are sets of pairs, for a relation R_s to include R means that $Graph(R) \subseteq Graph(R_s)$.

Furthermore, the smallest R_s to be symmetric and include R means that R_s is such that

$$(\forall r \in Relsr . Graph(R_s) \subseteq Graph(r)) \wedge R_s \in Relsr$$

where

$$Relsr = \{r | r \in Rel \wedge$$
$$Graph(R) \subseteq Graph(r) \wedge is\text{-}symmetric(r)\}$$

where *Rel* is the set of relations $A \times A \to \mathbb{B}$ to which R belongs, and of course

$$\text{is-symmetric }(R) \triangleq (\forall a, b \in A . aRb \Rightarrow bRa)$$

Definition 11.6

In general we can define closure with respect to any property of relations as follows. Let

$$R, R_c : A \times B \to \mathbb{B}$$

be relations defined from a set A to a set B with graphs

$$Graph(R), Graph(R_c) \subseteq A \times B$$

Let

$$P \in \mathcal{P}(A \times B) \to \mathbb{B}$$

be a predicate defined on the graphs of such relations. Then R_c is the *P-closure* of R if

(i) $Graph(R) \subseteq Graph(R_c)$

(ii) $P(Graph(R_c))$

(iii) $\forall R' \in A \times B \to \mathbb{B} . (P(Graph(R')) \Rightarrow Graph(R_c) \subseteq Graph(R'))$

Definition 11.7

Likewise, given $R \in A \times A \to \mathbb{B}$, the transitive closure R_T of R is such that

$$\forall r \in Reltr . Graph(R_T) \subseteq Graph(r) \wedge R_T \in Reltr$$

where

$$Reltr = \{r | r \in A \times A \to \mathbb{B} \wedge Graph(R) \subseteq Graph(r) \wedge \text{is-transitive }(r)\}$$

where

$$\text{is-transitive }(R) = (\forall a, b, c \in A . (aRb \wedge bRc) \Rightarrow aRc)$$

Exercise 11.1

Define a relation R_R which is the reflexive closure of a relation $R : A \times A \to \mathbb{B}$.

Example 11.5

A relation *Pred* is defined on the integers such that $m \, Pred \, n \Leftrightarrow m = n - 1$. The transitive closure of *Pred* is $Pred_T$ such that if $m \, Pred \, n$ then $m \, Pred_T \, n$ and if $m_1 \, Pred \, m_2 \wedge m_2 \, Pred \, m_3 \ldots m_x \, Pred \, n$ then $m_1 \, Pred_T \, n$. In fact $Pred_T = \, <$.

Exercise 11.2

What is the reflexive transitive closure of *Pred*?

Notation

Another way of defining a general closure often found in the literature is by means of a recursive definition; the reflexive transitive closure of $R: A \times A \to \mathbb{B}$ is for example R' where

(i) $\forall a, b \in A . aRb \Rightarrow aR'b$

(ii) $\forall a, b, c \in A . aR'b \wedge bR'c \Rightarrow aR'c$

(iii) $\forall a \in A . aR'a$

Further, R' must be the 'minimum' relation satisfying these criteria, that is that these are the only pairs $(a, b) \in A \times A$ which belong to R'. We allude here to a rather more complete and advanced theoretical treatment of recursion called the 'fixed point theory of recursion'. For a good treatment of fixed point theory and the application of minimum fixed points to recursion see either Bird (1976) or Stoy (1977).

Example 11.6

The reflexive, symmetric, transitive closure of any relation is itself a reflexive, transitive, symmetric relation and is therefore an equivalence relation. The reflexive, symmetric, transitive closure of $R \subseteq \mathbb{Z} \times \mathbb{Z}$ where $aRb \Leftrightarrow a - b = n$ is the relation of congruence modulo n.

Example 11.7: Product Numbers

With the product number suitable for computer programs we can define a relation 'supersedes' as follows: product number a supersedes b if a and b have the same product type, product number, and variant, and if the edition of a is greater than the edition of b. The reflexive symmetric closure of 'supersedes' is an equivalence relation.

An abstraction of this product-number example is to regard a product number as a tuple which is a member of the Cartesian product

$$Type \times Number \times Variant \times Medium \times Format \times Edition$$

We can define 'supersedes' more formally

let $\langle tu, na, va, ma, fa, ea \rangle \triangleq a$ **in**
let $\langle tb, nb, vb, mb, fb, eb \rangle \triangleq b$ **in**
a supersedes $b \triangleq ta = tb \wedge na = nb \wedge va = vb \wedge ea > eb$

11.3 PARTIAL ORDERS

We have encountered three standard properties of relations: symmetry, transivity, and reflexivity. Two other properties which are also useful are as follows

Definition 11.8
If R is a relation on A, R is *antisymmetric* if

$$\forall a, b \in A . aRb \wedge bRa \Rightarrow a = b$$

in other words R cannot relate two distinct elements of A to each other.

Definition 11.9
If R is a relation on A, R is *asymmetric* if

$$\sim (\forall a, b \in A . aRb \Rightarrow bRa)$$

In other words R is not symmetric: there is at least one pair of elements $a, b \in A$ such that aRb and $\sim bRa$.
  The more usually useful of these two properties is antisymmetry.

Partial Orders

Definition 11.10
A relation R on a set A is a *partial order* if it is reflexive, transitive and antisymmetric. The pair (A, R) is then called a *partially ordered set* or a *poset*.
  There are many examples of posets which occur in many different branches of mathematics.

Theorem 11.1

The relations $\leqslant$ and $\geqslant$ on the integers/natural numbers/real numbers are all partial orders. We merely verify the three properties.

Proof

$$a \leqslant a \qquad \text{(reflexivity)}$$
$$a \leqslant b \wedge b \leqslant a \Rightarrow a = b \qquad \text{(antisymmetry)}$$
$$a \leqslant b \wedge b \leqslant c \Rightarrow a \leqslant c \qquad \text{(transitivity)}$$

Likewise for $\geqslant$.

Example 11.8

In general, with a poset (A, R) we may have the situation where there are

$a, b \in A$ such that

   aRb

and

   bRa

so that R does not relate a and b in 'either direction'. This is not the case with any of the numeric examples with $\leqslant$ and $\geqslant$: given any a, b, either $a \leqslant b$ or $b \leqslant a$ for example. But given any family P of sets, the relation $\subseteq$ is a partial order. Consider

$$P = \{\{\ \}, \{a\}, \{b\}, \{c\}, \{b, c\}, \{a, c\}, \{a, b\}, \{a, b, c\}\}$$

We can take the familiar graph of the relation

	{ }	{a}	{b}	{c}	{b,c}	{a,c}	{a,b}	{a,b,c}
{ }	×	×	×	×	×	×	×	×
{a}		×				×	×	×
{b}			×		×		×	×
{c}				×	×	×		×
{b,c}					×			×
{a,c}						×		×
{a,b}							×	×
{a,b,c}								×

We can determine by inspection that the relation is reflexive: this is indicated by every position on the diagonal being occupied; the relation is antisymmetric: this is indicated by no two positions being occupied which are reflections of each other across the leading diagonal, except for positions on the diagonal itself; and it is transitive by inspection of each case.

This poset is an example of one in which there may be two members not related by the relation: for example

   $\sim \{a\} \subseteq \{b\}$

and

   $\sim \{b\} \subseteq \{a\}$

Exercise 11.3

Which of the following are partial orders? (Check whether each of the properties of reflexivity, transitivity and antisymmetry is present.)

(a) Given a set S, the relation $\supseteq$ defined by $a \supseteq b \triangleq b \subseteq a$.
(b) $<$ on the integers.

(c) $\Rightarrow$ on $\mathbb{B}$.

(d) $\Leftrightarrow$ on $\mathbb{B}$.

(e) $=$ on $\mathbb{Z}$.

(f) Given a set S, let M be the set of all total predicate functions $f:S\to\mathbb{B}$. Define a relation $\Rightarrow$ on M as follows

$$f\Rightarrow g \triangleq (\forall a\in S . f(a)\Rightarrow g(a))$$

Is $(M, \Rightarrow)$ a poset?

(g) Given a partial order R, define the inverse relation R^{-1}: $aR^{-1}b \triangleq bRa$. Is R^{-1} a partial order? Hence which of $\supseteq, \supset, \geqslant, >, \Rightarrow^{-1}$ are partial orders?

(f) In example (f), for each $f\in M$ we can define a particular relation $R_f:S\times S\to\mathbb{B}$

$$aR_f b \triangleq f(a)\Rightarrow f(b)$$

Is each R_f a partial order? Can you define a partial order on the set of relations

$$\{R_f|f\in M\}?$$

(i) In the last example of example 11.2 is 'supersedes' a partial order?

Notation: Diagrams

Partially ordered sets which are finite can be represented by diagrams. Each element of the set is represented by a node on the diagram, and a line connecting two nodes indicates that the lower of them is in relation to the upper. Since partial orders are by definition reflexive, the reflexive relation, that is the fact that every element is in relation to itself, is assumed and is not shown on the diagram. Likewise the transitive aspect of the relation is assumed and is not illustrated: if aRb and bRc, then the relationship aRc is not shown because it is deducible from the transitivity of partial orders. Hence in the diagram in figure 11.2 then the additional relation aRc (figure 11.3) is assumed and is not shown.

In figure 11.4 are some diagrams of certain partial orders. The second is isomorphic to the powerset of a set of three objects, with $\subseteq$. The third is isomorphic to four integers with $\leqslant$.

Figure 11.2

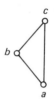

Figure 11.3

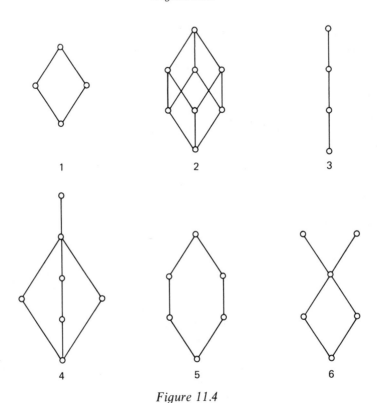

Figure 11.4

Exercise 11.4

Draw the second diagram in figure 11.4 labelling each node with one of the subsets of $\{a,b,c\}$.

Linear Orders

Definition 11.11
A *linear order* R on a set A is a partial order such that, for all $a,b \in A$,

either aRb or bRa; that is, a partial order with the property

$$\forall a, b \in A . aRb \lor bRa$$

Exercise 11.5

(i) Which of the partial orders found in exercise 11.3 are linear orders?
(ii) If R is a linear order, is R^{-1}?

Example 11.9: Product Numbers

In our product-number example, an engineering product (which could, for example, be a piece of software or a piece of electronic apparatus) may be composed of a number of parts, each of which in turn has a product number. Those products which are not composed of parts are 'atomic'. There results a relation 'is a part of', which can be interpreted as meaning that, if A is a part of B, then A is an immediate part of B.

We can further define a relation 'is a component of' which is the transitive closure of 'is a part of'. This means that, for example, if A is a part of B and B is a part of C, then A is a component of B, B is a component of C, and A is a component of C. We could further define a relation 'is found in' to be the reflexive transitive closure of 'is a part of', which would contain all the relationships in 'is a component of' and would also be reflexive, so that for all products A, A is found in A.

Thus given a collection of part numbers $A \ldots O$, a diagram illustrating 'is found in' might look as in figure 11.5. As shown in the diagram, everything is a component of A, and I is a component of B, C and A for example.

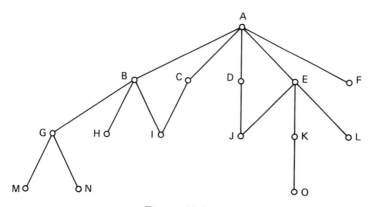

Figure 11.5

Exercise 11.6

As we have defined it, 'is found in' is reflexive and transitive. Bearing in mind that it is meant to represent a system of parts, is there any other property which this relation should have?

Notation

If R is a relation, the reflexive transitive closure of R is often written R^* and the transitive closure of R is sometimes written R^+.

11.4 LATTICES

Definition 11.12
Given a partially ordered set (A, R) and a subset $S \subseteq A$, then $a \in A$ is a *lower bound* of S if

$$\forall x \in S . aRx$$

This may seem intuitively clearer if the relation R is $\leqslant$. Then S has a lower bound a if

$$\forall x \in S . a \leqslant x$$

Definition 11.13
Similarly S has an *upper bound* $b \in A$ if

$$\forall x \in S . xRb$$

or

$$\forall x \in S . x \leqslant b$$

Notation

In these cases we may write:

$$aRS, \quad SRb$$

or

$$a \leqslant S, \quad S \leqslant b$$

respectively.

Example 11.10

Let

$$S = \{1, 2, 4, 8\}$$

then $1, 0, -1000$ are lower bounds and $8, 9, 10, 1000$ are upper bounds. We write

$$1 \leqslant S$$
$$-1000 \leqslant S$$
$$S \leqslant 8$$
$$S \leqslant 1000$$

etc.

Definition 11.14
a is called the *greatest lower bound* or *glb* of S if a is the greatest of all the lower bounds; that is, if

$$a \leqslant S \wedge \forall l \in A . l \leqslant S \Rightarrow l \leqslant a$$

or

$$aRS \wedge \forall l \in A . lRS \Rightarrow lRa$$

From now on in this section for clarity we will use $\leqslant$ as the symbol representing our relation.

We can write $\sqcap S$ for the glb of S.

Definition 11.15
Correspondingly, b is called the *least upper bound* or *lub* of S if b is the least of all the upper bounds; that is, if

$$S \leqslant b \wedge \forall u \in A . S \leqslant u \Rightarrow b \leqslant u$$

One may write $\sqcup S$ for the lub of S.

One particular subset of a poset A is A itself. Not all subsets of posets have lubs or glbs. For example, the integers with $\leqslant$ is a poset but the integers have no upper or lower bounds. Furthermore, certain subsets of the integers have no upper or lower bounds, such as the even integers, or the positive and negative primes.

Theorem 11.2

If a subset of a poset has a least upper bound, that lub is unique.

Proof
Let $S \subseteq (A, \leqslant)$ and let a, b be least bounds of S. Then

$$\forall x \in S . x \leqslant a \wedge x \leqslant b$$

Furthermore, since a and b are 'least'

$$\forall y \in A . S \leqslant y \Rightarrow a \leqslant y$$

Since $b \in A$ and $S \leqslant b$ then $a \leqslant b$.

Similarly $b \leqslant a$, but by the antisymmetry property of partial orders, if we have $a \leqslant b \wedge b \leqslant a$ then $a = b$. $\quad \square$

Corollary

There is a very similar proof that if a subset S of $(A, \leqslant)$ has a glb, then that glb is unique.

Example 11.11

In certain posets each subset has both a glb and a lub. An example is the set of all predicate functions defined on a set S, that is the set $S \rightarrow \mathbb{B}$, with the relation of $\Rightarrow$ as defined in exercise 11.3(f). Any two such functions $f, g : S \rightarrow \mathbb{B}$ have a glb and a lub called $f \sqcap g$, $f \sqcup g$ defined as follows

$$f \sqcap g(x) = f(x) \wedge g(x)$$
$$f \sqcup g(x) = f(x) \vee g(x)$$

Clearly

$$\forall x \in S . f \sqcap g(x) \Rightarrow f(x) \wedge f \sqcap g(x) \Rightarrow f(x)$$
$$\forall x \in S . f(x) \Rightarrow f \sqcup g(x) \wedge g(x) \Rightarrow f \sqcup g(x)$$

Notation

In general glbs and lubs may be regarded as operations and written $\sqcap$ and $\sqcup$ in infix notation. Thus $a \sqcap b$ is another way of writing $\sqcap \{a, b\}$, likewise for $\sqcup$.

Exercise 11.7

Prove that the operations of $\sqcap$, $\sqcup$, when they are defined on a poset, applied to two elements of that poset are associative and commutative. Show also that they idempotent; that is, that $a \sqcap a = a$, $a \sqcup a = a$.

Exercise 11.8

Draw a diagram of a finite poset having a subset which has no lub or glb and indicate the subset.

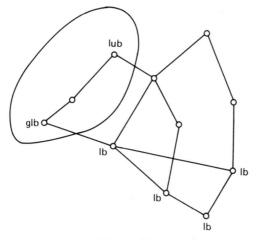

Figure 11.6

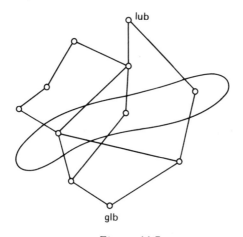

Figure 11.7

Example 11.12

The diagrams which are shown in figures 11.6 to 11.9 show subsets of posets which variously have upper bounds, least upper bounds, lower bounds, greatest lower bounds, etc. A subset may have an upper bound within itself or not within itself, and likewise for lower bounds.

Lattices

Definition 11.16
A partially ordered set in which every finite subset has a least upper bound and a greatest lower bound is called a *lattice*.

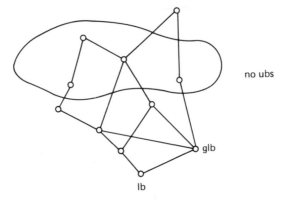

Figure 11.8

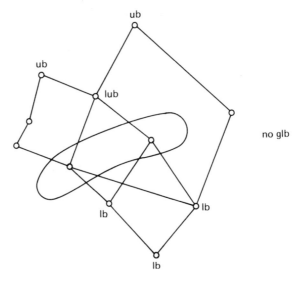

Figure 11.9

Theorem 11.3

It is sufficient that any pair of elements has a lub and a glb.

Proof
Consider any finite subset $S = \{s_1, s_2, \ldots, s_n\}$. Then the least upper bound of S, written $\sqcup S$, is, by the associativity of $\sqcup$ established in exercise 11.7,

$$S = \sqcup\{s_1, s_2, \ldots, s_{n-1}\} \sqcup s_n$$
$$= (((\ldots(s_1 \sqcup s_2) \sqcup s_3) \sqcup \ldots) \sqcup s_n$$

Thus we can see that if any pair of elements has a lub, then so does any finite subset. The same is of course true of glbs. A more rigorous proof uses induction over the cardinality of the subsets. ☐

Definition 11.17
A poset in which every subset (not just finite subsets) has a lub and a glb is called a *complete lattice*.

Examples 11.13

(i)   An example of a complete lattice is the powerset of a given set S with ⊆ as its order relation.
(ii) An example of a lattice which is not complete is the integers with ≤. Every finite set of integers has a maximum and a minimum, but infinite sets of integers, such as the multiples of 3 or the positive and negative primes, do not.
(iii) In the diagram in figure 11.10 we have added sufficient relations so that the poset as it was before is now a lattice. Readers may convince themselves of this.

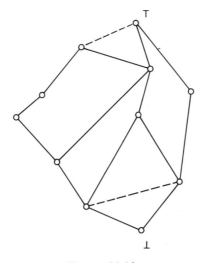

Figure 11.10

Notation

The element of a complete lattice which is the least upper bound of the whole lattice is called 'top' and is written '⊤'; the greatest lower bound is called 'bottom' and is written '⊥'.

Exercise 11.9

(a) Which of the posets found in exercise 11.3 are lattices?
(b) If (A, R) is a lattice, is (A, R^{-1}) a lattice? If (A, R) is complete, is (A, R^{-1}) complete?
(c) In the example at the end of section 11.3, is the set of product numbers with the reflexive closure of 'supersedes' a lattice? Is it complete?
(d) Is the set of parts, together with the relation 'is found in' as defined in exercise 11.6, a lattice? If it is not, what characteristics would the system have to have in order for it to be a lattice?

Notation

For a general lattice, in addition to '⊓' and '⊔' being used to indicate the glb and lub of two (or more) elements, for the relation itself the symbol '⊑' is often used. Thus, for the lattice $(\mathbb{Z}, \leqslant)$

⊑ would correspond to $\leqslant$

⊓ would correspond to Max

⊔ would correspond to Min

and for the lattice $(\mathbb{B}, \Rightarrow)$

⊑ would correspond to $\Rightarrow$

⊓ would correspond to $\land$

⊔ would correspond to $\lor$

11.5 APPLICATIONS

In this section we consider a number of examples of applications of the concept of lattices.

Example 11.14

Functions with a lattice as range.

Theorem 11.4

Functions with a lattice as range themselves form a lattice.

Proof
This is a generalisation of the observation we made earlier that the set of

functions

$$F = S \rightarrow \mathbb{B}$$

is a lattice, where if $f, g \in F$, $f \Rightarrow g$ is defined

$$f \Rightarrow g \triangleq (\forall x \in S . f(x) \Rightarrow g(x))$$

If we now have functions

$$G = S \rightarrow A$$

where L is a lattice $(A, \sqsubseteq)$, we can define $\sqsubseteq$ on members of G as follows

$$f \sqsubseteq g \triangleq (\forall x \in S . f(x) \sqsubseteq g(x))$$

The lub of f and g, $(f \sqcup g)$ is defined

$$(f \sqcup g)(x) \triangleq f(x) \sqcup g(x)$$

The property of being the least of all the upper bounds follows directly from $\sqcup$ being the least upper bound of two members of A.

Likewise the glb of f, $(f \sqcap g)$ is defined

$$(f \sqcap g)(x) = f(x) \sqcap g(x) \qquad \square$$

Definition 11.18
If, further, we consider functions which have a lattice as domain, and another lattice as range, say

$$h:L1 \rightarrow L2$$

such a function can have properties of a homomorphism in the sense of preserving the $\sqsubseteq$ relation. Such functions are called *monotonic*, as a generalisation of monotonic functions on the integers which preserve $\leqslant$ ($\leqslant$ being the lattice relation on the integers).

Definition 11.19
If in addition a function h preserves upper and lower bounds, it is said to be *continuous*. The notion of continuous functions on lattices plays an important part in constructing a model of the semantics of programming languages. See Stoy (1977) for a readable study in some depth of this topic.

Example 11.15: Specifications

Theorem 11.5

Given the functions of a certain type, $f:P \rightarrow Q$, we may consider the specifications of such functions f (recall chapter 10) which consists of pre-condition and post-condition

> **pre-**f:$P \rightarrow \mathbb{B}$
>
> **post-**f:$P \times Q \rightarrow \mathbb{B}$

These specifications form a lattice with respect to a 'naturally defined' ordering.

Proof

Given such a specification, we have a prescription for all functions which satisfy it

> $spec$:$(P \rightarrow Q) \rightarrow \mathbb{B}$
>
> $spec(f) = (\forall a \in P . \textbf{pre-}f(a) \Rightarrow \textbf{post-}f(a), f(a))$

Thus, for a given type of function f:$P \rightarrow Q$, we have specifications of type $(P \rightarrow Q) \rightarrow \mathbb{B}$.

Consider all such specifications

> $SPEC = (P \rightarrow Q) \rightarrow \mathbb{B}$

We can define an ordering $\Rightarrow$ as in the previous theorem, since $SPEC$ is a class of functions with $\mathbb{B}$ as a range. Thus

> $spec1 \Rightarrow spec2 \triangleq (\forall f \in P \rightarrow Q . spec1(f) \Rightarrow spec2(f))$

so any function which satisfies $spec1$ also satisfies $spec2$. This necessarily requires that the pre-conditions of $spec1$ are at least as broad as those of $spec2$; that is, that

> $\forall a \in P . pre2(a) \Rightarrow pre1(a)$

If this were not the case, we could find a value $b \in P$ such that

> $\sim (pre2(b) \Rightarrow pre1(b))$

that is

> $pre2(b) \wedge \sim pre1(b)$

This means that f can satisfy $spec1$ regardless of the value of $f(b)$, whereas $f(b)$ must satisfy $post2(b, f(b))$.

Any pair of specifications in $(P \rightarrow Q) \rightarrow \mathbb{B}$ has a lub and a glb: because they form a set of functions with $\mathbb{B}$ as a range, this follows from the previous subsection. The lub of $spec1$ and $spec2$ is

> $(spec1 \sqcup spec2)(f) \triangleq spec1(f) \vee spec2(f)$

and the glb is

> $(spec1 \sqcap spec2)(f) \triangleq spec1(f) \wedge spec2(f)$ $\square$

Note

The glb could easily be $\perp$ (bottom) of *SPEC* which cannot be satisfied by any function

$$spec_{\perp}(f) = \text{False}$$

Likewise there is the degenerately loose specification $\top$ (top) which any function f of the given type will satisfy

$$spec_{\top}(f) = \text{True}$$

A somewhat more thorough treatment of lattices applied to specifications of programs can be found in Denvir (1981).

Flow-diagrams

As an essay into the semantics of programming languages, Scott (1970) formulated flow-diagrams as a lattice. In this context a flow-diagram is somewhat akin to a path which control can take through a program. The referenced paper (Scott, 1970) is rewarding to read. It indicates the relevance of continuous functions to the fixed point theory of recursion (see also Bird (1976) and Stoy (1977)).

SUMMARY OF RESULTS

Result	Reference
$\geqslant$ and $\leqslant$ are partial orders	Theorem 11.1
lubs are unique	Theorem 11.2
If any pair of elements has a glb/lub, then so does any finite subset	Theorem 11.3
Functions with a lattice as range themselves are a lattice	Theorem 11.4
Specifications of functions of a given type form a lattice	Theorem 11.5

Chapter 12 Specification and Correctness of Programs

NOTATION INTRODUCED

Notation	Concept	Reference
wp	weakest pre-condition	Def. 12.1
$\{P1\}S\{P2\}$	proof rule	Section 12.3
$\dfrac{P1, P2, \ldots, Pn}{P}$	proof rule	Section 12.4

This chapter gives a brief survey of the issues of specification and correctness of programs. Other works referenced in section 12.7 deal with the subject in greater depth.

12.1 CORRECTNESS AND SPECIFICATIONS

We would like our programs to be correct. Furthermore we would like to be confident that they are correct and to impart that confidence to others. For example, the users, even the unconscious users of a program such as the passengers on an aeroplane which contains servo-systems or navigation equipment utilising embedded computers, would like to have confidence in the correct functioning of the software which may affect them. It may form part of a legal contract between supplier and client that the software supplied is of merchantable quality and performs to specification.

A program is an engineered construct: it is something the programmers have designed, built and delivered which is to perform a certain task. Programmers are designers in that they determine the nature and construction of the artifact which is supplied. The manner in which the artifact performs its function, the nature of the components (high level language statements) and the way they are composed are decided by the programmers. Programmers are therefore akin to designers of electronics circuits, rather than to those who simply wire up someone else's design. The latter correspond to computer operators who are instructed to make fifty copies of certain programs, perhaps instantiated with certain parameters for particular customisations, and place them in certain sequences on various physical volumes. In this respect, the role of a programmer is often underestimated.

A programmer, being a designer, has to ensure that the constructed program correctly performs the task for which it is intended; that is, that it meets its specification. Any engineer who designs a product (a bridge, aeroplane wing or VLSI circuit, for example) is expected to be able to defend the design produced with mathematically based arguments that it will meet its specification. Civil and electronic engineers model the behaviour of their designs, that is their putative artifacts, using mathematical models expressed in terms of continuous functions, calculus, the theory of complex variables and so forth. These modelling techniques form a 'phenomenological theory': there is a hypothesis that the mathematics expresses the real-world behaviour and the modification of that behaviour by the introduction of the putative artifact, adequately. By 'adequately' we mean that the deductions inferable from the model successfully predict details of the behaviour resulting from foreseeable events.

The reason why discrete mathematics is highly relevant to software engineering is that it supplies the notational equipment whereby we can express the behaviour of typical computer software, especially non-numeric programs such as are, with increasing frequency, resident in a computer embedded in some larger engineering product.

A specification from which deductions can be drawn, and about which reasoned arguments can be constructed, has to be a precise statement of the behaviour of the artifact in question. In this chapter we shall see how to use the notation described in this book to express the behaviour of a program, and hence to provide its specification. We shall also see how to prove that a given program satisfies a specification.

This approach is the basis of several recently developed techniques of program specification (Jones, 1980, 1986; Bjoerner and Jones, 1978; Bjoerner and Jones, 1982; Sufrin, 1981). In all of these, the concept of proving that a program correctly implements a specification is paramount.

'Verification' is the term given to the activity of proving a program correct. It has meaning only if there is a formal specification for the program. The specification has to be formal, that is expressed in a language with a precise, mathematically defined semantics, in order that statements about the program and its specification can be proved or disproved.

12.2 MODELS FOR PROGRAMS

Data

In order to write formal specifications for programs, we must have a mathematical semantics for the objects which programs manipulate. We shall start by providing a mathematical model of the 'virtual memory' of a program.

The 'virtual memory' of a high level language program consists of the collection of all the variables whose values that program can manipulate. The set of values which any given variable can adopt is its type. Thus, if we have a variable x declared

   x **Int**;

then the set of values which x may adopt is the set of all integers. (We ignore at least for the time being the question of limitations on the magnitude of integers representable in the computer.)

Example 12.1

If we have two integer variables

   x, y **Int**;

then the set of all values adoptable by x and y is the set of all pairs of integers, that is $\mathbb{Z} \times \mathbb{Z}$. If these were the only variables in the program then the 'state space' of the program would be $\mathbb{Z} \times \mathbb{Z}$. The various states which the program could be in at any time are

   $\langle -3, 76 \rangle$
   $\langle 4, 23 \rangle$
   $\langle 0, -1 \rangle$

etc.—that is, any pair of integers.

Thus in moving from a concrete machine containing addressable words, bytes, bits, to a high level language we move to a level of abstraction, and in considering the semantics of a high level language we move to a further level of abstraction.

Example 12.2

Suppose we have the following variables declared in a program

   x, y **Int**;
   a, b **String**;
   r **Real**;

We would then have a state space

$$\mathbb{Z} \times \mathbb{Z} \times \textbf{Char*} \times \textbf{Char*} \times \mathbb{R}$$

where Char is the set of all characters which may be used to form a 'string' in the programming language. At any given point in the execution of the program, the state of the program would consist of a tuple such as

$$\langle 37, -5, \text{"Press SPACE"}, \text{" "}, 3.14159 \rangle$$

This would constitute a 'snapshot' of the program state. In the activity of verification we use techniques for discussing such snapshots, relating them to each other, and making general statements about them.

Predicates

Predicates are used to constrain the state space to some sub-space.

Example 12.3

In example 12.2

$$x = 4$$

constrains the state space to

$$\{4\} \times \mathbb{Z} \times \textbf{Char*} \times \textbf{Char*} \times \mathbb{R}$$

which may be pictorially depicted as in fig. 12.1. This is difficult to draw since one is trying to represent a 5-dimensional space!

Any predicate function of the state space effectively defines a sub-space—the set of all those states for which the function is true.

Functions and Statements

Any statement, sequence of statements, procedure or program will carry out a transformation of the state. Thus, given any initial state, the statement etc. will produce a new state depending on it. Statements therefore map states to new states in a well-defined way, and are therefore associated with functions.

Example 12.4

(i) Consider the statement

$$x := x * x;$$

where the state is as before, that is $\mathbb{Z} \times \mathbb{Z} \times \textbf{Char*} \times \textbf{Char*} \times \mathbb{R}$. The

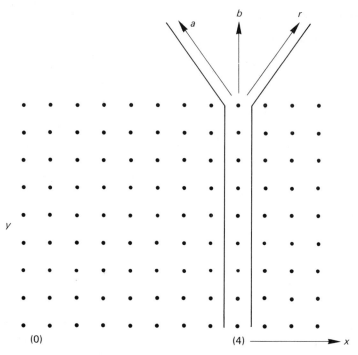

Figure 12.1

statement above transforms the state

$$\langle X, Y, S1, S2, R \rangle$$

where X is the value of the variable x, Y the value of y, etc., into

$$\langle X*X, Y, S1, S2, R \rangle$$

(ii) The statement

$$y := y + x - 2;$$

transforms

$$\langle X, Y, S1, S2, R \rangle$$

into

$$\langle X, Y + X - 2, S1, S2, R \rangle$$

(iii) The sequence of statements:

$$x := x*x;$$
$$y := y + x - 2;$$

transforms

$$\langle X, Y, S1, S2, R \rangle$$

into

$$\langle X*X, Y+(X*X)-2, S1, S2, R \rangle$$

Given that statements perform functions on the state space, the 'semantics' or meaning of statements can be expressed in terms of the specifications of their associated functions. We saw in chapter 10 that all functions can be specified by means of pre-conditions and post-conditions. This gives a strong clue to the way of specifying program statements. Pre-conditions and post-conditions play a large part in various methods of defining program language semantics.

12.3 WEAKEST PRE-CONDITIONS

Definition 12.1
Given any statement S in a program, if R is a predicate function of the value space which is required to be true after the statement is executed, then there is a weakest pre-condition which, if true before the execution of the statement, will ensure that R is true after.

Example 12.5

Let us take an example. Consider again the statement

$$x := x * x;$$

and let

$$R \triangleq (x = 4)$$

Then for R to hold, that is for $x = 4$ to be true after the statement, the necessary pre-condition is

$$(x = -2) \vee (x = 2)$$

There are other, stronger, conditions which will also ensure the truth of R after S. For example

$$x = 2$$

or

$$(x = 2) \wedge (y = 7)$$

If conditions C and D are such that $C \Rightarrow D$ we say that C is *stronger* than

D, and D is *weaker* than C. Thus in the above

$$x = 2$$

is stronger than

$$(x = -2) \vee (x = 2)$$

and

$$(x = 2) \wedge (y = 7)$$

is stronger than

$$(x = -2) \vee (x = 2)$$

because

$$(x = 2) \Rightarrow ((x = -2) \vee (x = 2))$$

and

$$((x = 2) \wedge (y = 7)) \Rightarrow ((x = -2) \vee (x = 2))$$

If P is the weakest pre-condition then any other pre-condition Q which ensures that R is true after executing S is such that

$$Q \Rightarrow P$$

Notation

We write

$$P = \text{wp}\,(S, R)$$

or, in the case of example 12.5

$$((x = -2) \vee (x = 2)) = \text{wp}\,(x := x * x, x = 4)$$

Theorem 12.1

Given a statement S and a post-condition R, of all the pre-conditions of S ensuring the post-condition R, there is a unique weakest one.

Proof
The proof lies in the fact that all such pre-conditions are members of $V \rightarrow \mathbb{B}$ where V is the state space, and in chapter 11 we saw that functions $V \rightarrow \mathbb{B}$ with $\Rightarrow$ form a complete lattice. This means that any subset has a least upper bound and a greatest lower bound. In this context it means that each subset, such as the set of all pre-conditions ensuring R to be true after executing S, has a strongest and a weakest member. The strongest pre-condition is 'False' (because 'False' implies any statement), and the weakest is the one we are interested in. $\square$

Interpretation

The weakest pre-condition is the condition which must hold in order that the post-condition holds after the statement. It may be impossible for the post-condition to be true. For example, for $x = -4$ after $x := x * x$ is executed, where x is of type integer, we would find that the weakest pre-condition is False.

Notation

Another notation is to write

$$\{(x = -2) \lor (x = 2)\} x := x * x \{x = 4\}$$

Assignments

Example 12.6

In the foregoing example, our required post-condition was

$$x = 4$$

For this to be true after

$$x := x * x$$

it is necessary that

$$(x * x) = 4$$

before the assignment. However

$$(x * x) = 4$$

is simply another way of writing

$$(x = -2) \lor (x = 2)$$

This is an instance of the general rule for assignments.

Rule
If we wish $R(x)$ to be true after the assignment

$$x := e$$

then the weakest pre-condition is

$$R(e)$$

that is the predicate $R(x)$ rewritten with the expression e standing in place of all the free occurrences of x. Thus

$$\{R(e)\} x := e \{R(x)\}$$

or alternatively

$$R(e) = \text{wp}\,(x:=e, R(x))$$

Exercise 12.1

(a) What is wp $(x:=x*x, y=4)$?
(b) What is wp $(x:=x*x, y=4 \wedge x=9)$?
(c) What is wp $(y:=y+x-2, y=4 \wedge x=9)$?

It is of course possible that there is no pre-condition such that the post-condition will hold after executing the statement.

Example 12.7

Given that x is Real

$$\text{wp}\,(x:=x*x, x=-3)$$

is, if we apply the rule

$$x*x=-3$$

There is no Real value of x which will satisfy the above, and so this is False for all Real values of x; hence it is equivalent to

False

In other words, the post-condition is impossible to satisfy.

Example 12.8

The opposite situation can arise. Consider

$$\text{wp}\,(x:=3, x=3)$$

This works out as

$$3=3$$

which is equivalent to

True

Exercise 12.2

(a) What is wp $(y:=y+x-2, y=x+2)$?
(b) What is wp $(y:=x-2, y=x+2)$?
(c) What is wp $(y:=x-2, y=2-x)$?
(d) What is wp $(y:=x-2, y=x-2)$?

Interpretation

When the weakest pre-condition is 'True' it means that regardless of the value of the state-space before the statement is executed, the post-condition will hold afterwards. If we find that the weakest pre-condition for a statement S and post-condition R is True, then we have proved that R holds after S.

If we have a specification for a function $f:A \rightarrow B$ consisting of

pre-$f:A \rightarrow B$

post-$f:A \times B \rightarrow \mathbb{B}$

a program S which is to satisfy the specification for f should be such that

pre-$f(a) \Rightarrow \text{wp}(S, \textbf{post-}f(b))$

where a and b are the values of the variables which model the input argument and result respectively of f.

Example 12.9

Thus if f is specified

$f:\mathbb{R} \rightarrow \mathbb{R}$

pre-$f(x) \triangleq \text{True}$

post-$f(x, r) \triangleq r = x^2 + 2x + 1$

the following single statement program will fulfil the specification for f

$r := x*x + 2*x + 1$

To prove this we have to show that

pre-$f(x) \Rightarrow \text{wp}(r := x*x + 2*x + 1, \textbf{post-}f(x, r))$

that is

$\text{True} \Rightarrow \text{wp}(r := x*x + 2*x + 1, r = x*x + 2*x + 1)$

that is

$\text{wp}(r := x*x + 2*x + 1, r = x*x + 2*x + 1)$

Using the rule for assignment, this is

$x*x + 2*x + 1 = x*x + 2*x + 1$

that is

True

Which proves that the program fulfils the specification. The proof looks laborious and obvious, and in practice we would take suitable short cuts to reduce the working.

Sequences

Programs which consist of single assignment statements are of course extremely limited. We can define the weakest pre-conditions, however, for all the other structured program constructs. Sequences are straightforward. Suppose we have

$$S_1; S_2$$

with a required post-condition R. Then, for R to hold after executing S_2, $wp(S_2, R)$ must hold before it. However, if $wp(S_2, R)$ holds before executing S_2, it must hold immediately after executing S_1. Thus, before executing S_1, we must have

$$wp(S_1, wp(S_2, R))$$

Rule

If therefore we have a sequence

$$S_1; S_2; \ldots S_n$$

at the end of which Rn must hold, then we can as it were 'pull the condition back' through the sequence

$$\{R_0\}S_1\{R_1\}S_2\{R_2\}\ldots\{R_{n-1}\}S_n\{R_n\}$$

where for each i

$$R_i = wp(S_{i+1}, R_{i+1})$$

Example 12.10

> average: $\mathbb{R} \times \mathbb{R} \times \mathbb{R} \rightarrow \mathbb{R}$
> pre-average $(a, b, c) = \text{True}$
> post-average $(a, b, c, r) = 3 * r = a + b + c$

The program P is

> $r := a;$
> $r := r + b;$
> $r := r + c;$
> $r := r/3$

Then to prove P fulfils average, we have to show that

$$\text{True} \Rightarrow \text{wp}\,(P, 3*r = a+b+c)$$

that is

$$\text{wp}\,(r:=a; r:=r+b; r:=r+c,$$
$$\text{wp}\,(r:=r/3,\, 3*r=a+b+c)) \Leftrightarrow$$
$$\text{wp}\,(r:=a;\, r:=r+b,\, \text{wp}\,(r:=r+c,\, r=a+b+c)) \Leftrightarrow$$
$$\text{wp}\,(r:=a,\, \text{wp}\,(r:=r+b,\, r+c=a+b+c)) \Leftrightarrow$$
$$\text{wp}\,(r:=a,\, \text{wp}\,(r:=r+b,\, r=a+b)) \Leftrightarrow$$
$$\text{wp}\,(r:=a,\, r+b=a+b) \Leftrightarrow$$
$$\text{wp}\,(r:=a,\, r=a) \Leftrightarrow$$
$$a=a \Leftrightarrow$$
$$\text{True}$$

A useful technique is to decorate the program with comments in a progressive way, to form the proof. Thus

(i)

$$r:=a;$$
$$r:=r+b;$$
$$r:=r+c;$$
$$r:=r/3$$
$$\{3*r=a+b+c\}$$

(ii)

$$r:=a;$$
$$r:=r+b;$$
$$r:=r+c;$$
$$\{r=a+b+c\}$$
$$r:=r/3$$
$$\{3*r=a+b+c\}$$

until the final stage one obtains

$$\{\text{True}\}$$
$$r:=a;$$
$$\{r=a\}$$
$$r:=r+b;$$

$$\{r = a + b\}$$

$$r := r + c;$$

$$\{r = a + b + c\}$$

$$r := r/3$$

$$\{3 * r = a + b + c\}$$

Exercise 12.3

Play the following game with two colleagues. First choose a few variables of suitable types. Then ask your two colleagues to invent a short sequence (4 or 5) of 'random' assignment statements using the variables. Finally ask for a predicate again using only the variables chosen. Taking this predicate as the post-condition of the sequence, establish the weakest pre-condition for the sequence.

12.4 AXIOMATIC SEMANTICS

Notation

We can postulate rules for the ways in which assertions, in the form of predicates on the state-space, are transformed by executing all the forms of statement in a language; or we can postulate rules of inference, whereby given a predicate in relation to one or more types of statement, we can infer other predicates. Such rules form the 'Axiomatic Semantics' for the language.

These Axiomatic Semantics take two forms, which are very strongly related. They are

(i) Predicate Transformers
This is the form we have mostly shown so far, in which a rule is given for deriving the weakest pre-condition from a given post-condition. Recall that the pre-conditions and post-conditions constrain the state-space, so that the weakest pre-condition, constituting the weakest constraint, corresponds to the largest subset of values of the state-space each of which will ensure the post-condition.
(ii) Proof Rules
These assert a relation between the pre-condition and post-condition. For composite statements these assertions will be in terms of assertions about the components.

Assignments

Notation: Predicate Transformers

We have seen the Predicate Transformer form of this rule

$$\text{wp}\,(x := e, P(x)) = P(e)$$

For completeness, we should add the rule for a null statement

$$\text{wp}\,(\text{NULL}, P) = P$$

Thus NULL does not change any assertions about the state.

Notation: Proof Rules

The corresponding proof rule forms are, for assignments

$$\{P(e)\}x := e\{P(x)\}$$

and for NULL

$$\{P\}\,\text{NULL}\,\{P\}$$

Let us elaborate on this format. The normal form of a proof rule is

Assertion 1, Assertion 2, Assertion n
———————————————————
Assertion x

meaning that one may infer assertion x from assertions 1 to n. Any free variables occurring in these may be instantiated with any values of the appropriate type, provided that all occurrences are instantiated with the same value. Thus, for example above, we can instantiate P with any predicate on the state-space

$$\{x = 4\}\text{NULL}\{x = 4\}$$

or

$$\{e = y + 7\}x := e\{x = y + 7\}$$

or

$$\{a + b = y + 7\}z := a + b\{z = y + 7\}$$

etc.

Interpretation

If there is no horizontal line, and just one assertion in the proof rule, as is the case for assignments and NULL, we can infer the rule (or a suitable instantiation of it) without any previous assertions. These, the rules about assignments and NULL, are our starting points.

The interpretation of a form

$$\{P\}S\{Q\}$$

is to be read: if P holds before executing S then Q will hold after it. In general if a program P is to model a function f, we wish to prove that

$$\{\textbf{pre-}f\}P\{\textbf{post-}f\}$$

Sequences

Notation: Predicate Transformer

We have been introduced to the Predicate Transformer rule for sequences. It is

$$\text{wp}\,(S_1;S_2,P)=\text{wp}\,(S_1,\text{wp}\,(S_2;P))$$

Notation: Proof Rule

We have also had an intimation of the proof-rule form. It is

$$\frac{\{P\}S_1\{Q\},\,\{Q\}S_2\{R\}}{\{P\}S_1;S_2\{R\}}$$

Applying this rule iteratively, we can deduce

$$\frac{\{P_0\}S_1\{P_1\},\,\{P_1\}S_2\{P_2\},\ldots,\{P_{(n-1)}\}S_n\{P_n\}}{\{P_0\}S_1;S_2;\ldots;S_n\{P_n\}}$$

In our example of the average of three numbers a,b,c, the assertions between each statement correspond to the predicate P_0, P_1 etc.

Conditional Statements

Assignments and sequences are very limited forms of statement. We need a few other constructs, one of which is the conditional statement. This can take the form

if B **then** S_1 **else** S_2

Notation: Predicate Transformer

The Predicate Transformer rule is

$$\text{wp}\,(\textbf{if }B\textbf{ then }S_1\textbf{ else }S_2,P)\triangleq$$
$$(B\wedge\text{wp}\,(S_1,P))\vee(\sim B\wedge\text{wp}\,(S_2,P))$$

Thus, if B is true, the weakest pre-condition is that of S_1, and if B is not true it is that of S_2.

Notation: Proof Rule

The proof rule for the same type of statement is

$$\frac{\{P \wedge B\}S_1\{Q\}, \{P \wedge \sim B\}S_2\{Q\}}{\{P\} \text{ if } B \text{ then } S_1 \text{ else } S_2\{Q\}}$$

All the usual forms of conditional statement can be reconstructed in terms of the simple form shown. For example

    if B **then** S_1

is equivalent to

    if B **then** S_1 **else** NULL

and

    case $B_1 : S_1$
        $B_2 : S_2$
        $\vdots$
        $B_n : S_n$
        else: $S_{(n+1)}$ **esac**

is equivalent to

    if B_1 **then** S_1 **else if** B_2 **then** S_2 **else**...
        else if B_n **then else** $S_{(n+1)}$

Example 12.11

Let us take a very simple example of a program which delivers the maximum of two values. The program is as follows

    if $a \geqslant b$ **then** result$:=a$ **else** result$:=b$

The post-condition (which looks very like the program itself) is

$$a \geqslant b \Rightarrow result = a \wedge b \geqslant a \Rightarrow result = b$$

The similarity is deceptive for the above is a purely logical assertion, whereas the program is a statement which is executed. Let P stand for the post-condition above, and S stand for the program. Then

$\text{wp}\,(S, P) =$

$(B \wedge \text{wp}\,(\text{result}:=a, P)) \vee (\sim B \wedge \text{wp}\,(\text{result}:=b, P)) =$

$(B \wedge (a \geqslant b \Rightarrow \text{True} \wedge b \geqslant a \Rightarrow a = b)) \vee$

$(\sim B \wedge (a \geqslant b \Rightarrow b = a \wedge b \geqslant a \Rightarrow \text{True})) =$

$(a \geqslant b) \vee (b \geqslant a) = \text{True}$

which proves that the program satisfies the required post-condition.
   The proof rule leads to the following argument

$$\{a \geqslant b \wedge (b \geqslant a \Rightarrow a = b)\}\text{result}:=a\{P\} \tag{1}$$

Simplifying (1)

$$\{a \geqslant b\}\text{result}:=a\{P\} \tag{2}$$

$$\{a \geqslant b \Rightarrow b = a \wedge b \geqslant a\}\text{result}:=b\{P\} \tag{3}$$

Simplifying (3)

$$\{b \geqslant a\}\text{result}:=b\{P\} \tag{4}$$

$$\{b \geqslant a\}\text{result}:=a\{P\} \tag{5}$$

$$\{a \geqslant b\}\text{result}:=a\{P\}, \{b \geqslant a\}\text{result}:=\{P\} \tag{6}$$

Applying *Modus Ponens* (see below) in (6) from (2) and (5)

$$\{\text{True}\}\ \textbf{if}\ a \geqslant b\ \textbf{then}\ \text{result}:=a\ \textbf{else}\ \text{result}:=b\{P\} \tag{7}$$

which proves the result.

Proofs

A few notes on general principles of applying proof rules are relevant at
this point. First, the principle of *Modus Ponens* means that if one has the
following as given facts

$$P_1, P_1, \ldots, P_n$$
$$\frac{P_1, P_2, \ldots, P_n}{P_x}$$

one may then infer

$$P_x$$

A second principle is that given

$$\{Q\}S\{R\}$$

one may infer

$$\{P\}S\{R\}$$

where $P \Rightarrow Q$.

Iterations

The third main program language construct which one needs for producing structured programs is that of iteration. A suitable form of this is

While B **do** S **od**

Notation: Predicate Transformers

The semantics of this construct are more difficult to define than those of the other constructs. The predicate transformer form is

wp (**While** B **do** S **od**, R) $= (\exists i \in \mathbb{N}_0 . H_i)$

where

$$H_0 = R \wedge {\sim} B$$

and for $i \geqslant 0$

$$H_{(i+1)} = B \wedge \text{wp}(S, H_i)$$

that is

$$H_1 = B \wedge \text{wp}(S, H_0)$$
$$H_2 = B \wedge \text{wp}(S, H_1)$$

etc. This means that there must be some $i \geqslant 0$ such that H_i is true before executing the loop for R to be true afterwards. Then the loop will execute i times.

Notation: Proof Rule

The proof rule is

$$\frac{\{P \wedge {\sim} B \Rightarrow R\}, \{P \wedge B\} S \{P\}}{\{P\} \textbf{ while } B \textbf{ do } S \{R\}}$$

However, this proof rule ensures only partial correctness: there is no guarantee that the loop will terminate. If it does terminate however, R will hold after it is executed. The Predicate Transformer form does ensure termination however, since it states

$$(\exists i \in \mathbb{N}_0 . H_i)$$

some finite i exists such that the loop terminates after i iterations.

The diagram in figure 12.2 illustrates the 'Invariance Theorem'. This is strongly related to the proof rule and states the following.

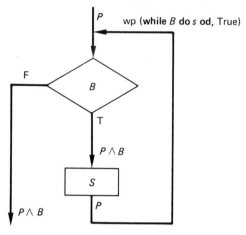

Figure 12.2

Theorem 12.2

Let P be such that

$$(P \wedge B) \Rightarrow \mathrm{wp}\,(S, P)$$

so that if P and B hold before executing S, P will hold after S. Then

$$(P \wedge \mathrm{wp}\,(\textbf{while } B \textbf{ do } S \textbf{ od}, \mathrm{True})) \Rightarrow$$
$$\mathrm{wp}\,(\textbf{while } B \textbf{ do } S \textbf{ od}, P \wedge \sim B)$$

Proof (sketch)
P is the same P as in the proof rule. Since

$$\{\mathrm{wp}\,(S, P)\}S\{P\}$$

then

$$\{P \wedge B\}S\{P\}$$

The clause

$$\mathrm{wp}\,(\textbf{while } B \textbf{ do } S \textbf{ od}, \mathrm{True})$$

is the weakest pre-condition which will ensure that the loop will terminate (It is, if one thinks about it, the weakest pre-condition such as to ensure the weakest possible post-condition; that is, that 'anything will be true' after executing the statement, or that the statement has a 'hereafter'!)

Illustration

In figure 12.2, P is the 'invariant'. It is 'invariantly true' throughout the execution of the loop: that is, S preserves the truth of P. If P and B hold before S, P will hold after. This means that one can be sure that if P holds before the loop, P will still hold afterwards.

Proving Termination

The proof of termination is always difficult. One approach is to find a bounded integer function which is monotonic with respect to each passage through the loop. Thus if $f(v)$, where v is the value of the state-space, is such that $f(v) < max$ and the execution of S causes $f(v)$ to increase, then we can deduce termination of the loop. For if before the execution of the loop $f(v) = f_0$, then the loop can be executed a maximum of $(max - f_0)$ times before it terminates.

Example 12.12

When constructing a loop, or looking for a proof of correctness of a loop, a search for an invariant is usually very fruitful. If we recall exercise 6.3, the invariant is the parity of the number of black balls. This is not altered by the 'move' and so is the same before starting the loop and after finishing it. Since we terminate with one ball, this is black if we start with an odd number of black balls, and white if we start with an even number. The number of balls in the bag is the function of the state-space, which is bounded, and monotonic: it decreases by one with each move.

Procedures and Functions

Rule

A procedure body is a sequence which has its own specification in terms of its pre-conditions and post-conditions, which would be parameterised with the values of the arguments of the procedure. In that case the body of the procedure can be proved to satisfy its specification in the normal way, and a call to that procedure may have as its weakest pre-condition that given by the pre-condition and post-condition of its body.

Example 12.13

> Proc Square (x, y)
> $y := x * x$ end;

Then

> $\text{wp}(\text{Square}(x, y), R(x, y)) =$
> $\text{wp}(y := x * x, R(x, y)) =$
> $R(x, x * x)$

which gives the formula for the weakest pre-condition for the call of the procedure Square(x, y).

If we express the same thing as a function

Function Square(x);

it should have been designed to fulfil a specific pre-condition and post-condition; in that case we can infer a proof rule

$$\frac{\textbf{pre-}Square(x)}{\textbf{post-}Square(x, r)}$$

In this case it is likely that the pre-condition and post-condition are

pre-$Square(x) \triangleq$ True

post-$Square(x, r) \triangleq r = x * x$

which means that one may infer

$Square(x) = x * x$

when manipulating logical expressions.

Example 12.14

Our square-root (x, t) was an example of a function with both pre-condition and post-condition. If we have

$x := squareroot(x, t) + y$

then the weakest pre-condition would be

wp $(x := squareroot(x, t) + y, R(x, t, y)) =$

pre-$squareroot(x, t) \wedge (\textbf{post-}squareroot(x, t, r) \Rightarrow R(r + y, t, y))$

which in this case is

$x \geqslant 0 \wedge t > 0 \wedge (|r * r - x| \leqslant t \Rightarrow R(r + y, t, y))$

Exercise 12.4

Given the above, what is

(i)   wp $(x := squareroot(x, t) + y, x = 4 + y)$,

(ii)   wp $(x := squareroot(x, t), x \geqslant 3 \wedge x \leqslant 4)$.

Arrays

Arrays affect our proof rules because they can stand on the left-hand side of assignments. The normal rule for assignments applies in principle, but care has to be taken because of possible 'aliasing'.

Example 12.15

If

$$R \triangleq A(i) = 0$$

and we have

$$A(j) := A(j) - 1$$

then

$$\text{wp} \, (A(j) := A(j) - 1, R) =$$
$$i = j \Rightarrow A(j) = 1 \, \wedge$$
$$i \neq j \Rightarrow A(i) = 0$$

12.5 CONSEQUENCES FOR SOFTWARE ENGINEERING

The Materials We Use

When setting out to build any engineered artifact, we prefer to use materials whose behaviour we believe we understand well; that is, whose behaviour we can describe with a reasonably simple mathematical model, from which it is easy to predict the actual behaviour when it is placed in any of the usual contexts or environments. Thus, for example, electrical engineers tend to use metallic conductors, semiconductors, or insulators (or materials which behave as insulators in a particular well-defined environment). They tend not to use organic materials, liquid solutions, mixed gases etc. Civil engineers prefer to use materials whose strength does not degrade with the passing of time—materials whose performance can be defined and relied upon.

In the foregoing section we have defined the behaviour of a number of software engineering materials: assignments, sequences, iterations, conditionals, procedures, and functions. We may include blocks or compound statements, since these are equivalent to in-line procedures. We did not include 'go to' statements, and we were somewhat cautionary about aliasing of array elements. This is because the definition of the behaviour of 'go to' statements is extremely awkward and inferring conclusions from a context containing 'go to' statements or their destinations is very difficult. The equations defining the behaviour of a sequence, for example, no longer hold if a 'go to' can transfer control into the middle of the sequence.

The equations and rules governing the behaviour of program language statements are the 'Ohm's and Kirchoff's laws' of programming; just as Ohm's and Kirchoff's laws do not hold for non-metallic conductors or conductors mounted on imperfect insulators, so our rules of program statements do not apply if we use materials outside the collection whose behaviour we have defined.

This in a nutshell is the rationale behind the basic precepts of structured programming and the 'go to considered harmful' dictum (Dijkstra, 1968).

Constructive Design

To take a program which is already written, and prove that it is correct, that is, to prove that it meets a given specification, is a feasible but arduous task. The approach to constructing a design, as indeed in other engineering activities, is to design according to the criteria which will progressively be met in order that the design will satisfy the specifications. In other words, the designer designs bearing in mind that the design will have to be 'defended', that it will be required to demonstrate that the design can be proved to meet the specification.

This process is indeed carried out in more traditional disciplines. Any civil engineer would expect to be able to defend the assertion that his design for an airport runway could bear the specified weight of aircraft, the electrician to defend that his plan for house wiring would support the calculated current load distributed among the outlets. My first employment was as a young apprentice in the application group of a semiconductor manufacturer. In those days a single silicon transistor of the cheapest type cost the equivalent of three days' pay; the most expensive were equivalent to three weeks' pay. Hence my section leader was concerned to ensure that my designs were correct before I built and tested them. He required me to convince him that my design as expressed on paper 'would work', and when I had done so to his satisfaction, he would agree to my building it.

The 'manufacture' of software, that is the submitting of a program to a computer and initial trial of it, is extremely cheap, by contrast. This heavily mitigates against a discipline of ensuring that designs will 'work' in advance. The tendency which results is to construct a program rapidly and try it out as soon as possible. One is then immediately operating in a mode of 'tinkering' with the program until it passes the tests which one has thought of. This is of course no guarantee that the program is correct: the tests may be more or less adequate, but they can never be complete, except for a very simple program. In general, tests can show the presence of errors, but never their absence.

Hence the philosophy we would advocate is that of 'constructive design', that is designing according to a proof. The previous sections have given some intimations of this, and the principles are admirably explained and exposed by Jones (1980, 1986). A superb sequence of examples of this process is given in Dijkstra (1976), in which he develops programs in the constructive style by producing a sequence of assertions. The spaces in between each assertion are progressively filled with programming statements which can be proved to fulfil the assertion following them, treated as a post-condition, given the previous assertion, treated as a pre-condition. This constructive technique is finely illustrated by Dijkstra (1976) in a sequence of examples of increasing sophistication. The basis of constructive design is on proof techniques which depend on an axiomatic semantics of programming language statements (Hoare, 1969).

Validation and Scientific Method

The distinction between a specification and a design of a piece of software is not that well defined. While a specification defines what is to be done rather than how it is to be achieved in algorithmic terms, the data may be formulated in terms of particular models, and/or the total process may be decomposed into a number of sub-processes, the behaviour of each of which is defined in the specification. The process of design is step-wise: parts of the specification are 'reified' or made more concrete, and parts are decomposed into further processes which are specified. Jones (1980, 1986) refers to this process as the 'Rigorous Method'.

At any stage in the design process, the designer should be prepared to show that the latest stage of the design satisfies the requirements imposed by the previous stage. This activity is often termed 'validation', and when the design is implemented the usual form of the validation activity is that of testing.

Whether it takes the form of testing or proof, the activity of validation is an experiment. The designer is proposing a hypothesis that his design meets its requirements. This hypothesis is akin to a scientific theory: the next step is to design and carry out experiments whose general aim is to increase confidence in the validity of the theory. The appropriate way of doing this is to design experiments to refute the hypothesis. Thus, when carrying out proofs of correctness, if not proving total correctness of the whole design, the choice of which parts to prove should be those which are most suspect, critical, and about which one's peer designers/analysts feel least convinced. If the validation consists of testing, the choice of tests should concentrate on the exceptional situations: the boundary values and pathological cases and combinations of them. This is fundamental to Scientific Method, and can be applied to engineering design activities also.

The same ideas can guide us when debugging a program. We are then in a situation where our 'theory' has been proved incorrect by the failure of some test. The notions of proof techniques can guide us in deciding what to do next in order to isolate the fault. Suppose the program is represented as in figure 12.3. We are in the position where we thought we

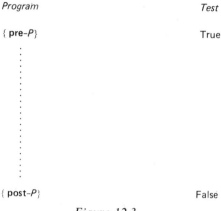

Figure 12.3

could prove the program P correct if called upon to do so, but this has been refuted by a test case. The next step is to divide the program P into parts such that we feel sure that it is feasible to prove the correctness of each part, and where the proof of the program as a whole follows immediately from the correctness of the parts. This is illustrated in figure 12.4. Here Assertion 1 acts as a post-condition for Part 1 and a pre-condition for Part 2. If we insert a check for Assertion 1 into the program we shall be able to determine whether Part 1 or Part 2 is at fault. Then the process is repeated until we are faced with a piece of program of manageable size whether either the error is apparent or the proof can be carried out statement by statement and the discrepancy found.

Of course, programs do not usually consist of straight sequences of statements, but may consist of conditions or loops as shown in section 12.4. The proof may be divided according to the diagrams in figures 12.5 and 12.6. Checks for the extra assertions are inserted in the program, and the failing test re-run. In the diagram in figure 12.6 Assertion 1 is devised so that it is an invariant of the loop, and so that

(i)   **pre-***P* $\Rightarrow$ Assertion 1
(ii)   (Assertion 1 $\wedge \sim B) \Rightarrow$ **post-***P*

It is good engineering practice to insert such checks in the program when writing it in the first place. This is facilitated by languages such as CHILL and Ada which have specific ASSERT statements. In other languages a standard procedure may be devised for the purpose.

This technique of proof is known as 'proof by sub-goal assertions'. Each assertion is a sub-goal on the path to proving the goal, which is the correctness of the whole program.

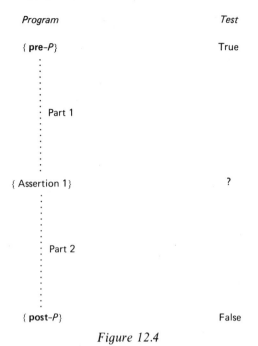

Figure 12.4

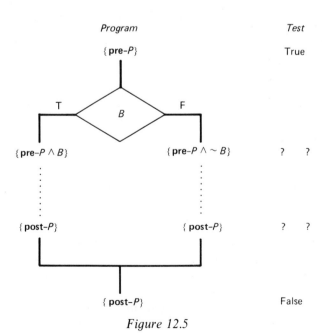

Figure 12.5

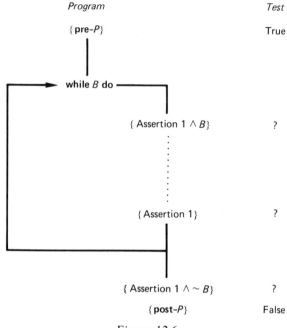

Figure 12.6

An Engineering Discipline

The principles outlined in this chapter, that is the use of suitable 'programming materials', constructive design, and applying a scientific method to both the design process and to fault-finding, constitute the basis of an engineering discipline for software development. For the last few years many writers of repute have been exhorting those in the practice of software engineering to adopt a more disciplined approach based on the foregoing notions. See, for example, Jones (1986), Mills (1980), Gries (1980) and Hindlater and Shapiro (1981).

12.6 WORKSHOP EXERCISES

In the following, if exercises 9.19(1) and 10.6 have been tackled, it may be more appropriate to use these as the specification for the algorithms and proofs you are asked to construct in exercises 12.5 to 12.8.

Exercise 12.5

Referring back to both exercises 8.12 and 10.5(1), for the space-borne computer system operations were specified and their schemes defined, namely the operations which calculate the future position, future velocity etc. For one or more of the operations so specified, produce an algorithm (which may invoke other functions specified but not designed) and walk through an outline of its proof. If a suitable compiler is available, insert checks of intermediate assertions and run some tests through the checks.

Exercise 12.6

Referring back to exercises 8.13 and 10.5(2), for the fault-tolerant control system select one or more of the operators for which schemes and specifications were previously written. Produce an algorithm and walk through a proof as indicated in exercise 12.5 above.

Exercise 12.7

Referring back to the personal diary problem in exercises 8.14 and 10.5(3), select one or more operations whose schemes and specifications were defined (adding, deleting, or moving an event or any other operation defined). Proceed as indicated in exercise 12.5 above.

Exercise 12.8

Referring back to the nuclear power station problem in exercises 8.15 and 10.5(4), select one or more of the operations whose schemes and specifications were defined and proceed as indicated in exercise 12.5 above.

12.7 FURTHER READING

Software Engineering

Jones (1986) and Bjoerner and Jones (1982) carry the application of discrete mathematics to the formulation of a method for developing software systems, called the Vienna Development Method (VDM). This is a good, practical, formally based approach to software development, which uses some of the discrete mathematics covered in this book as models for algebras in software specifications. In fact, the notation in parts of this book, especially in chapter 10, has been biased towards that used in VDM so as to ease the step which any interested reader would have to take in order to assimilate VDM, and I therefore acknowledge again the inspiration which the latter has provided.

A more abstract approach to specifications can be found in the work of Burstall and Goguen (1980a, 1980b) and the ADJ group (Goguen *et al.*, 1977). This work is the basis for more abstract specification methods and may lead to practical software engineering methods in the future.

Algebra

Those interested in reading more about algebra for the fascination intrinsic to the subject itself are recommended to peruse and absorb MacLane and Birkhoff (1979). The algebraic approach to specifications depends on Category Theory for its theoretical underpinning, for which the introductory text written by Krishnan (1981) or Arbib and Manes (1975) is recommended; however, a grasp of abstract algebra and topology is necessary in order to assimilate these works.

Computer Science

Discrete Mathematics has traditionally been the basis of computer science theory rather than software engineering; it is its relation to the latter which has been the whole motivation of this book. However, an excellent work for those who wish to explore computer science aspects is Beckman (1980).

Happy further reading!

SUMMARY OF RESULTS

Result	Reference
Weakest pre-conditions are unique	Theorem 12.1
Axioms (predicate transformers and proof rules) for NULL statements, assignments, sequences, conditions, and iterations	Section 12.4
Invariance Theorem	Theorem 12.2
Proof of termination	Section 12.4
Proof rules for procedures, functions and arrays	Section 12.4
Debugging by sub-goal assertions	Section 12.5

References

Arbib, M.A. and Manes, E.G. (1975). *Arrows, Structures and Functors*, Academic Press.

Backhouse, R.C. (1979). *Syntax of Programming Languages, Theory and Practice*, Prentice-Hall.

Beckman, F.S. (1980). *Mathematical Foundations of Programming*, Addison-Wesley.

Bird, R. (1976). *Programs and Machines*, Wiley.

Bjoerner, D. and Jones, C.B. (1978). 'The Vienna Development Method: The Meta-Language', *Lecture Notes in Computer Science, Vol. 61*, Springer-Verlag.

Bjoerner, D. and Jones, C.B. (1982). *Formal Specification and Software Development*, Prentice-Hall.

Burstall, R.M. and Goguen, J.A. (1980a). *An informal introduction to specifications using CLEAR*, Computer Science Dept., University of Edinburgh.

Burstall, R.M. and Goguen, J.A. (1980b). 'The Semantics of CLEAR, a Specification Language', *Lecture Notes in Computer Science, Vol. 86: Abstract Software Specifications*, Springer-Verlag.

Chomsky, N. (1956). 'Three Models for the Description of Language', *IRE Trans. Information Theory*, **IT-2**, p. 3.

Denvir, B.T. (1981). 'A Lattice-Theoretic Approach to System Specifications and the Contractual Methodology', *STL Telecommunications Network Report 18*.

Dijkstra, E.W. (1968). 'Go To Statement Considered Harmful', *Communication ACM*, **11**, pp. 147–8.

Dijkstra, E.W. (1976). *A Discipline of Programming*, Prentice-Hall.

Gödel, K. (1931). 'Über formal unentscheidbare Sätze der Principia Mathematica und verwandter Systeme I', *Monatschafte für Mathematik und Physic*, **38**, pp. 175–98.

Goguen, J.A., Thatcher, J.W., Wagner, E.G. and Wright, J.B. (1977). 'Initial Algebra Semantics and Continuous Algebras', *Journal of the ACM*, **24:1**, pp. 68–95.

Gries, D. (1980). 'Educating the Programmer: Notations, Proofs and the Development of Programs', *IFIP80*, North-Holland.

Hindlater, R. and Shapiro, S.D. (1981). 'A Program of Continuing Education in Applied Computer Science', *IEEE Computer*, October 1981.

Hoare, C.A.R. (1969). 'The Axiomatic Basis of Computer Programming', *Communication ACM*, **12**, pp. 576–83.

Jones, C.B. (1980). *Software Development: A Rigorous Approach*, Prentice-Hall.

Jones, C.B. (1986). *Systematie Software Development Using VDM*, Prentice-Hall.

Krishnan, V.S. (1981). *An Introduction to Category Theory*, North-Holland.

MacLane, S. and Birkhoff, G. (1979). *Algebra*, 2nd edn, Collier–Macmillan.

Mills, H. (1980). *Structured Programming in Pascal*, University of Maryland.

Milner, R. (1980). 'A Calculus of Communicating Systems', *Lecture Notes in Computer Science*, *Vol. 92*, Springer-Verlag.

Nagel, E. and Newman, J.R. (1959). *Gödel's Proof*, Routledge and Kegan Paul.

Quine, W.V.O. (1980). *Elementary Logic*, Harvard University Press.

Scott, D. (1970). 'The Lattice of Flow Diagrams', *Symposium on Semantics of Algorithmic Languages, Lecture Notes in Mathematics 188*, Springer-Verlag.

Stoy, J.E. (1977). *Denotational Semantics*, MIT Press.

Sufrin, B. (1981). 'Formal Specification of a Display Editor', *PRG-21*, Oxford University Programming Research Group.

Tennant, N.W. (1978). *Natural Logic*, Edinburgh University Press.

Answers to Exercises

PART I

Chapter 2

Exercise 2.1

(a) $\{2, 4, 6, 8, 10\}$
(b) $\{a, b, c\}$

Exercise 2.2

$A, B; C, D$

Exercise 2.3

(a), (c), (d)

Exercise 2.4

(a) $\{0, -1, 1, -2, 2, -3, 3\}$
(b) $\{3\}$
(c) $\{x \mid x \in \mathbb{Z} \text{ and } x \bmod 5 = 0\}$

Chapter 3

Exercise 3.1

x	y	$x \wedge y$	$y \wedge x$	$x \vee y$	$y \vee x$	$x \Leftrightarrow y$	$y \Leftrightarrow x$	$x \Rightarrow y$	$y \Rightarrow x$
T	T	T	T	T	T	T	T	T	T
T	F	F	F	T	T	F	F	F	T
F	T	F	F	T	T	F	F	T	F
F	F	F	F	F	F	T	T	T	T

Exercise 3.2

x	y	z	$x \wedge y$	$(x \wedge y) \wedge z$	$y \wedge z$	$x \wedge (y \wedge z)$
T	T	T	T	T	T	T
T	T	F	T	F	F	F
T	F	T	F	F	F	F
T	F	F	F	F	F	F
F	T	T	F	F	T	F
F	T	F	F	F	F	F
F	F	T	F	F	F	F
F	F	F	F	F	F	F

x	y	z	$x \vee y$	$(x \vee y) \vee z$	$y \vee z$	$x \vee (y \vee z)$
T	T	T	T	T	T	T
T	T	F	T	T	T	T
T	F	T	T	T	T	T
T	F	F	T	T	F	T
F	T	T	T	T	T	T
F	T	F	T	T	T	T
F	F	T	F	T	T	T
F	F	F	F	F	F	F

x	y	z	$x \Leftrightarrow y$	$(x \Leftrightarrow y) \Leftrightarrow z$	$y \Leftrightarrow z$	$x \Leftrightarrow (y \Leftrightarrow z)$
T	T	T	T	T	T	T
T	T	F	T	F	F	F
T	F	T	F	F	F	F
T	F	F	F	T	T	T
F	T	T	F	F	T	F
F	T	F	F	T	F	T
F	F	T	T	T	F	T
F	F	F	T	F	T	F

Consider

x	y	z	$x{\Rightarrow}y$	$(x{\Rightarrow}y){\Rightarrow}z$	$y{\Rightarrow}z$	$x{\Rightarrow}(y{\Rightarrow}z)$
T	T	T	T	T	T	T
T	T	F	T	F	F	F
T	F	T	F	T	T	T
T	F	F	F	T	T	T
F	T	T	T	T	T	T
F	T	F	T	F	F	T
F	F	T	T	T	T	T
F	F	F	T	F	T	T

Thus $(F{\Rightarrow}F){\Rightarrow}F$ is false whereas $F{\Rightarrow}(F{\Rightarrow}F)$ is true, so $\Rightarrow$ is not associative.

Exercise 3.3

(a)

x	y	z	$y \wedge z$	$x \vee (y \wedge z)$	$x \vee y$	$x \vee z$	$(x \vee y) \wedge (x \vee z)$
T	T	T	T	T	T	T	T
T	T	F	F	T	T	T	T
T	F	T	F	T	T	T	T
T	F	F	F	T	T	T	T
F	T	T	T	T	T	T	T
F	T	F	F	F	T	F	F
F	F	T	F	F	F	T	F
F	F	F	F	F	F	F	F

(b)

x	y	z	$y \vee z$	$x \wedge (y \vee z)$	$x \wedge y$	$x \wedge z$	$(x \wedge y) \vee (x \wedge z)$
T	T	T	T	T	T	T	T
T	T	F	T	T	T	F	T
T	F	T	T	T	F	T	T
T	F	F	F	F	F	F	F
F	T	T	T	F	F	F	F
F	T	F	T	F	F	F	F
F	F	T	T	F	F	F	F
F	F	F	F	F	F	F	F

Exercise 3.4

(a)

x	$\sim x$	$\sim(\sim x)$
T	F	T
F	T	F

(b)

x	y	$x \Leftrightarrow y$	$x \Rightarrow y$	$y \Rightarrow x$	$(x \Rightarrow y) \wedge (y \Rightarrow x)$
T	T	T	T	T	T
T	F	F	F	T	F
F	T	F	T	F	F
F	F	T	T	T	T

Exercise 3.5

(b)   Let $G \triangleq$ Fred's house is green
let $R \triangleq$ Fred's house is red
then $G \vee R$

(c)   Let B be the set of London buses
let $G(x) \triangleq x$ is painted green
let $R(x) \triangleq x$ is painted red
then $x \in B \Rightarrow (G(x) \vee R(x))$

(d)   Let $R \triangleq$ it is raining
let $P \triangleq$ it is pouring
then $R \Rightarrow P$
or equivalently
$\sim (R \wedge \sim P)$

(e)   Let $C \triangleq$ my car is on a cobbled road
let $F \triangleq$ my car is going fast
then $C \Rightarrow (\sim F)$

(f)   Let S be the set of students
let $C(x) \triangleq x$ takes computer science
let $E(x) \triangleq x$ takes electronic engineering
then $x \in S \Rightarrow ((C(x) \vee E(x)) \wedge \sim (C(x) \wedge E(x)))$

Exercise 3.6

(a) Let $b \triangleq$ set of London buses
let $R(x) \triangleq x$ is red
let $G(x) \triangleq x$ is green
then $\forall x \in b . R(x) \vee G(x)$

(b) Let $C(x) \triangleq x$ is on a cobbled road
let $F(x) \triangleq x$ is going fast
let $age(x)$ be the age of x in years
let V be the set of cars
then $\forall x \in V . age(x) > 20 \Rightarrow \sim (C(x) \wedge F(x))$

(c) Let H be the set of houses in our street
let $R(x) \triangleq x$ is red
let $G(x) \triangleq x$ is green
then $\forall x \in H . R(x) \vee G(x)$

(d) Let B be the set of London buses
let $R(x) \triangleq x$ is red
let $G(x) \triangleq x$ is green
then $\exists x \in B . R(x) \wedge \exists x \in B . G(x)$

Chapter 4

Exercise 4.1

(a) 0
(b) 1
(c) $\{0\}$
(d) $\{0, \{0\}\}$
(e) $\{0, \{0\}, \{\{0\}\}, \{0, \{0\}\}\}$
(f) 2^{n-1}

Exercise 4.2

(a) Verifying Boolean Algebra axioms for the $\mathbb{B}$ model:
(i) $x \wedge y = y \wedge x$
for proof see answer to exercise 3.1.
(ii) $(x \wedge y) \wedge z = x \wedge (y \wedge z)$
for proof see answer to exercise 3.2.

(iii) $z \wedge \sim z = F$. If $x \wedge \sim y = F$ then either $x = F$ or $y = T$ or both. In these cases (see below) $x \wedge y = x$.

	x	y	$\sim y$	$x \wedge \sim y$	$x \wedge y$
*	T	T	F	F	T
	T	F	T	T	F
*	F	T	F	F	F
*	F	F	T	F	F

(iv) $z \wedge \sim z = F$.

If $x \wedge y = x$ then $x \wedge \sim y = F = z \wedge \sim z$ (see above).

(b) Verifying the Boolean axioms for $S = \{a, b, c\}$:

(i) Inspecting the table for $\cap$ shows a symmetry about the diagonal demonstrating commutativity.

(ii) The associativity $x \cap (y \cap z) = (x \cap y) \cap z$ can be checked for all values S1–S8 of x, y, z by inspection. For example

$$(S1 \cap S2) \cap S3 = S2 \cap S3 = S4$$

$$S1 \cap (S2 \cap S3) = S1 \cap S4 = S4$$

(iii) Inspection of the tables for $\cap$ and ' shows that for all values S1–S8 of z, $z \cap z' = S8$.

For the cases where $x \cap y' = S8$,

either (a) $x = S8$

or (b) $y' = S8$ in which case $y = S1$

or (c) $y' = x'$—that is, $y = x$

or (d) six other individual cases.

Dealing with each of these in turn, by inspection of the tables,

(a) if $x = S8$, $x \cap y = S8 = x$ for any y

(b) if $y' = S8$, $y = S1$ and $x \cap S1 = x$ for any x

(c) if $y = x$ then $x \cap y = x \cap x = x$

(d) the last six cases can be summarised in the following table

x	y'	y	$x \cap y$
S6	S4	S5	S6
S4	S6	S3	S4
S4	S7	S2	S4
S7	S4	S5	S7
S6	S7	S2	S6
S7	S6	S3	S7

showing that in each case $x \cap y = x$.

(iv) The converse argument to (iii) also holds since each step can be reversed.

PART II

Chapter 5

Exercise 5.1

$$D^+ \times \{.\} \times D^* \times \{E\} \times \{+, -\} \times D^+$$

where

$$D = \{0, 1, 2, 3, 4, 5, 6, 7, 8, 9\}$$

Exercise 5.2

$$A \times \overset{5}{\underset{n=0}{\oplus}} (A \oplus D)^n$$

alternatively

$$\underset{n=0}{\oplus} A \times (A \oplus D)^n$$

Identifiers of any length

$$A \times (A \oplus D)^*$$

Exercise 5.3

Let $Sign \triangleq \{`+', `-'\}$.

(i) $Sign \times (D^+)$

Let $Point \triangleq \{`.'\}$.

Let $Mant \triangleq ((D^* \times Point) \oplus (Point \oplus D^+) \oplus (D^+ \times Point \times D^+))$

(ii) $Sign \times Mant$
(iii) $Sign \times Mant \times \{`E'\} \times ((Sign \times D^+) \oplus D^+)$

Exercise 5.4

(i) b
(ii) 3
(iv) $\langle b \rangle$
(v) a
(vi) $\langle b, c, c, b, a \rangle$

(vii) $\langle b, c \rangle$
(viii) c
(ix) d

Exercise 5.5

len dels(S, n) = (len S) − 1
dels(S, n) is defined if and only if $0 < n \leqslant$ len S
len mods(S, n, e) = len S
mods(S, n, e) is defined if and only if $0 < n \leqslant$ len S
dels(S, n) = dels mods(S, n, e)
mods$(S, n, S(n))$ = S
len subs(S, n, k) = $k + 1 \geqslant 1$
subs$(S, 1, \text{len}(S) - 1)$ = S
subs(S, n, k) is defined if and only if $n + k \leqslant$ len S

There are no doubt other relationships to be found.

Chapter 6

Exercise 6.1

   Events $(Rooms \times P) \rightarrow (T \times C \times L)$

This is the strict answer to the question. Although not more than one lesson can take place in a given room at once, the same lesson can be associated with more than one room at the same time, as we have prescribed it above.

Exercise 6.2

(i)   If the teachers can have free periods, then the mapping

        $P \rightarrow (C \times L)$

        which is the range of the *Teacher-schedule* mapping, is partial.
(ii)   If classes can have free periods, then the mapping

        $P \rightarrow (T \times L)$

        which is the range of the *Class-schedule* mapping, is partial.
(iii)   If rooms can be sometimes empty, *Events* is partial.

Exercise 6.3

In each of the three possible moves, one either removes two black balls or zero black balls: for one either:

(a) removes two white balls, replacing one, which removes no black balls; or

(b) removes two black balls, and places a white one back, which removes two black balls; or

(c) removes a black and a white ball and returns a black ball, which removes no net black balls.

Hence all possible moves preserve the parity (even-ness or odd-ness) of the black balls in the bag. Hence the algorithm is a parity checker: if there are an odd number of black balls in the bag to start with, one finishes with a black ball, otherwise with a white ball.

Exercise 6.4

> subs: $A^* \times \mathbb{N} \times \mathbb{N}_0 \to A^*$
> subs$(l, n, k) \triangleq$ **if** $n > $ len l **then** $\langle \ \rangle$ **else**
> > **if** $n = 1$ **then** front$(l, k - 1)$
> > **else** subs$(\text{tl } l, n - 1, k)$

Exercise 6.5

> mods: $A^* \times \mathbb{N} \times A \to A^*$
> **pre**-mods$(l, n, a) \triangleq n \leqslant$ len l
> mods$(l, n, a) \triangleq$
> > **if** $n = 1$ **then** $\langle a \rangle \| \text{tl } l$
> > **else** $\langle \text{hd } l \rangle \| \text{mods}(\text{tl } l, \ n - 1, a)$
> subs: $A^* \times \mathbb{N} \times \mathbb{N}_0 \to A^*$
> **pre**-subs$(l, n, k) \triangleq n + k \leqslant$ len l
> subs$(l, n, k) \triangleq$
> > **if** $n = 1$ **then** front$(l, k - 1)$
> > **else** subs$(\text{tl } l, n - 1, k)$

Chapter 7

Exercise 7.1

See figures A.7.1–A.7.3.

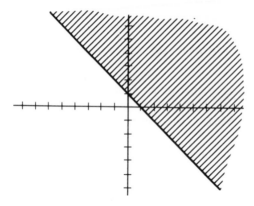

Figure A.7.1

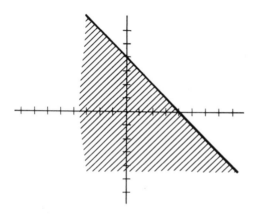

Figure A.7.2

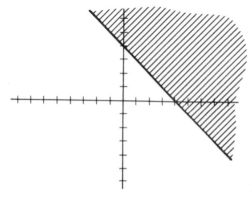

Figure A.7.3

Exercise 7.2

$4 \leqslant x^2 + y^2 \leqslant 9$

Exercise 7.3

$f(x) \triangleq \{ y | 4 \leqslant x^2 + y^2 \leqslant 9 \}$

Exercise 7.4

(i)   Transitive and symmetric. Not reflexive, since one is not normally regarded as one's own sibling.
(ii)   Transitive.
(iii) Transitive.
(iv)   None.
(v)   Transitive.
(vi)   Symmetric.
(vii) Symmetric.
(viii) Transitive.
(ix)   Symmetric.

PART III

Chapter 8

Exercise 8.1

$\{ \}$, because for any X, $\{ \} \cup X = X \cup \{ \} = X$

Exercise 8.2

X, because for any subset of X, $S \cap X = X \cap S = S$

Exercise 8.3

No, for if I were an identity, then

$I - 2 = 2$ so $I = 4$
$I - 3 = 3$ so $I = 6$

Exercise 8.4

No: there is no truth value X such that

$\text{False} \Rightarrow X \equiv \text{False}$

Exercise 8.5

$+, *, \cup, \cap, \vee, \wedge$

Exercise 8.6

(i)   No, for there is no truth value X such that

$$\text{False} \wedge X = \text{True}$$

(ii)   No, for there is no truth value X such that

$$\text{True} \vee X = \text{False}$$

Exercise 8.7

(i)   The integers with $+$ is a group.
(ii)   The integers with $-$ is none of these because $-$ is not associative.
(iii)   The integers with $*$ is a monoid.
(iv)   The integers with $/$ is none of these.
(v)   The reals with $+$ is a group.
        The reals with $-$ is none of these.
        The reals with $*$ is a monoid, for 0 has no inverse.
        (The reals without 0 with $*$ is a group.)
        The reals with $/$ is none of these.
(vi)   The Booleans with $\vee$ are a monoid.
        The Booleans with $\wedge$ are a monoid.
        The Booleans with $\Rightarrow$ are none of these, for $\Rightarrow$ is not associative.
(vii) $\{ \ \}$ is an identity for $\cup$, but if $x \in S$, S being non-empty, there is no
        inverse T of $\{x\}$ such that $T \cup \{x\} = \{ \ \}$. So $\mathscr{P}S$ with $\cup$ is a monoid.
        Likewise $\mathscr{P}S$ with $\cap$ is a monoid.

Exercise 8.8

(i)   A ring.
(ii)   Neither.
(iii)   A field.
(iv)   Neither.
(v)   Neither: $\mathbb{B}$ is not a group with respect to $\wedge$ because F has no
        inverse.
(vi)   If n is prime, then the integers mod n is a field, otherwise a ring. We
        show first that they form a ring.

$$\text{Let } \mathbb{Z}_n \triangleq \text{integers mod } n = \{x \mid x \in \mathbb{Z} \wedge 0 \leqslant x < n\}$$
$$\text{Define } +_n : x +_n y \triangleq (x + y) \bmod n$$
$$\text{Define } *_n : x *_n y \triangleq (x * y) \bmod n$$

$+_n$ has identity 0, is associative, and every $x \in \mathbb{Z}_n$ has an inverse $(n-x)$ in $\mathbb{Z}_n$. So $(\mathbb{Z}_n, +_n, 0)$ is a group.

$*_n$ has identity 1, and is associative.

We prove the distributive rule as follows. For $x, y, z \in \mathbb{Z}_n$

$$
\begin{aligned}
x *_n (y +_n z) &= (x * ((y+z) \bmod n)) \bmod n \\
&= (x * (y+z)) \bmod n \text{ if } y+z < n \\
&= (x * (y+z) + x * n) \bmod n \text{ if } y+z \geqslant n \\
&= (x * (y+z)) \bmod n \text{ in either case} \\
&= (x * y + x * z) \bmod n \\
&= (x * y) +_n ((x * z) \bmod n) \\
&= (x *_n y) +_n (x *_n z)
\end{aligned}
$$

The other distributive rule follows from this one by commutativity of $*_n$ and $+_n$.

Hence $\mathbb{Z}_n$ with $+_n$ and $*_n$ is a ring.

To show that if n is prime, $\mathbb{Z}_n$ is a field, suppose: $x, y, z, \in \mathbb{Z}_n$ and $x *_n y = x *_n z$. Then

$$(x *_n y) +_n (-_n (x *_n z)) = 0$$

so, by distributivity, since $\mathbb{Z}_n$ is a ring

$$(x *_n y) +_n (x *_n (-_n z)) = 0$$

that is

$$x *_n (y +_n (-_n z)) = 0$$

that is

$$x * (y +_n (-_n z)) = k * n \text{ for some } k$$

Since $x, y, z \in \mathbb{Z}_n$

$$0 \leqslant x < n \quad \text{and} \quad 0 \leqslant (y +_n (-_n z)) < n$$

Since n is prime and cannot divide x or $(y +_n (-_n z))$, either $(y +_n (-_n z)) = 0$ or $x = 0$. But we have shown that $\mathbb{Z}_n$ with $+_n$ is a group, and elements of groups have unique inverses. So if $(y +_n (-_n z)) = 0$ then $y = z$.

Therefore for every distinct y, z and non-zero x in $\mathbb{Z}_n$, $x *_n y$ and $x *_n z$ are distinct.

There are n different values which can be assigned to $y \in \mathbb{Z}_n$, and therefore n possible values of $x *_n y$, of which one must be 1, the identity for $*_n$. Hence there is some y in $\mathbb{Z}_n$ such that $x *_n y = 1$, so every non-zero x has an inverse with respect to $*_n$. Thus if n is prime $\mathbb{Z}_n$ is a field.

Exercise 8.9

(i)   In a ring $(R, +, *)$

$$0*b + a*b = (0 + a)*b \qquad\qquad \text{by the distributive law}$$
$$= a*b = 0 + a*b$$

$(R, +, 0)$ is a group, so by the cancellation law

$$0*b = 0$$

(ii)   by the above
$$0 = 0*(-b)$$
$$= (a + (-a))*(-b)$$
$$= a*(-b) + (-a)*(-b) \qquad\qquad \text{by the distributive law. Also}$$
$$0 = a*0 \qquad\qquad\qquad\qquad\qquad \text{by the above}$$
$$= a*((-b) + b)$$
$$= a*(-b) + a*b \qquad\qquad \text{by the distributive law.}$$

Cancelling on the left

$$a*b = (-a)*(-b)$$

Exercise 8.10

$$0*a = 0$$
$$(1 + (-1))*a = 0$$
$$1*a + (-1)*a = 0$$
$$a + (-1)*a = 0$$
$$(-1)*a = -a$$

Exercise 8.11

$A \times B$ is a group with respect to $+$.
$+$ is associative. Proof:

$$(a, k) + ((b, l) + (c, m)) =$$
$$(a, k) + (b + c, l + m) =$$
$$(a + (b + c), k + (l + m)) =$$
$$((a + b) + c, (k + l) + m) =$$
$$(a + b, k + l) + (c, m) =$$
$$((a, k) + (b, l)) + (c, m)$$

(0,0) is an identity for $+$ and the inverse of (a, b) is $(-a, -b)$.

$*$ is similarly associative.

$*$ is distributive with respect to $+$. Proof:

$$(a, k) * ((b, l) + (c, m)) =$$
$$(a, k) * (b + c, l + m) =$$
$$(a * (b + c), k * (l + m)) =$$
$$(a * b + a * c, k * l + k * m) =$$
$$(a * b, k * l) + (a * c, k * m) =$$
$$(a, k) * (b, l) + (a, k) * (c, m)$$

A similar argument demonstrates the other distributive rule. Thus $A \times B$ with $*$, $+$ is a ring.

In general if A, B are fields, $A \times B$ is not a field because although $(1, 1)$ is an identity with respect to $*$, $(0, a)$, $a \neq 0$ is not an identity with respect to $+$ and $(0, a)$ has no inverse with respect to $*$.

Chapter 9

Exercise 9.1

Let $a, b \in A$. Then $a * b = b * a$.

But $h(a) * h(b) = h(a * b) = h(b * a) = h(b) * h(a)$

thus $*$ is commutative with respect to at least those members of B which are images of A under h.

Exercise 9.2

Let $a, b, c \in R$. Then:
$$a * (b + c) = a * b + a * c$$

But $h(a) * (h(b) + h(c)) =$
$$h(a) * h(b + c) =$$
$$h(a * (b + c)) =$$
$$h(a * b + a * c) =$$
$$h(a * b) + h(a * c) =$$
$$h(a) * h(b) + h(a) * h(c)$$

Hence the distributive rule applies to those members of S which are images of R under h. The other distributive rule similarly applies.

Exercise 9.3

$f(x+y)=(x+y) \bmod n$

$(f(x)+f(y)) \bmod n = ((x \bmod n)+(y \bmod n)) \bmod n$

Let $x=k*n+x \bmod n$

Let $y=l*n+y \bmod n$

then $x+y=(k+l)*n+x \bmod n+y \bmod n$

and $(x+y) \bmod n = (x \bmod n+y \bmod n) \bmod n$

therefore $f(x+y)=(f(x)+f(y)) \bmod n$

Exercise 9.4

Consider all the pairs of members of A with their unions:

$\{\ \}\cup\{a\}=\{a\}$; $\{\ \}\cup\{a,b\}=\{a,b\}$; $\{a\}\cup\{a,b\}=\{a,b\}$

their images under h are:

$\{\ \}\cup\{\ \}=\{\ \}$ in each case. The same holds for intersection.

If we now consider B with union and intersection:

$\{\ \}\cup\{\ \}=\{\ \}$; $\{\ \}\cap\{\ \}=\{\ \}$

The images under g are again:

$\{\ \}\cup\{\ \}=\{\ \}$; $\{\ \}\cap\{\ \}=\{\ \}$; being members of A.

Exercise 9.5

Epimorphism:

$h:b\rightarrow A$

$h(\{\ \})=\{\ \}, h(\{a\})=\{\ \}$

Monomorphism:

$f:A\rightarrow B$

$f(\{\ \})=\{\ \}$

Isomorphism:

$g:B\rightarrow C$

$g(\{\ \})=\{\ \}, g(\{a\})=\{b\}$

There is another homomorphism $j:A\rightarrow B$

$j(\{\ \})=\{a\}$

Many others can be found.

Exercise 9.6

$m(\Rightarrow)$:

	$m(T)=\{\ \}$	$m(F)=\{x\}$
$m(T)=\{\ \}$	$T=m^{-1}\{\ \}$	$F=m^{-1}\{x\}$
$m(F)=\{x\}$	$T=m^{-1}\{\ \}$	$T=m^{-1}\{\ \}$

Thus the induced operation is

	$\{\ \}$	$\{x\}$
$\{\ \}$	$\{\ \}$	$\{x\}$
$\{x\}$	$\{\ \}$	$\{\ \}$

that is, the inverse of '$-$', so that $a\,m(\Rightarrow)\,b\equiv b-a$

Exercise 9.7

$=$ maps to $'$ (complement)
$\Rightarrow$ maps to $-$ (difference)

Exercise 9.8

$h(p+a)=h(p)+h(a)=0_G+h(a)=h(a)$
$h(a+p)=h(a)+h(p)=h(a)+0_G=h(a)$

$m(\langle x,y\rangle+\langle p,q\rangle)=$
$m(\langle x+p,y+q\rangle)=$
$\langle ax+ap+by+bq,\,Ax+Ap+By+Bq\rangle=$
$\langle ax+by,\,Ax+By\rangle+\langle ap+bq,\,Ap+Bq\rangle=$
$m(\langle x,y\rangle)+m(\langle p,q\rangle)$

Exercise 9.10

If $aB=Ab$ the plane is mapped onto a line, and m is not an isomorphism; for example, if $A=B=0$, $a=b=1$, the plane is mapped onto the x-axis. Otherwise m is an isomorphism; for example, if $a=B=1$, $b=A=0$, m is the identity mapping.

Exercise 9.11

Let $h:A \rightarrow B, k:B \rightarrow C$

(i)   For any $x \in C$ we can find $y \in B$ such that $k(y) = x$. For any $y \in B$ we
      can find $z \in A$ such that $h(z) = y$. Hence for any $x \in C$ we can find $z \in A$
      such that $k(h(z)) = x$ so $k_o h$ is an epimorphism.
(ii) For any $x \in C$ if $k(y) = x$ and $k(z) = x$ then $y = z$. For any $y \in B$ if
      $h(a) = h(b)$ then $a = b$. Hence for any $x \in C$ if $k(h(a)) = k(h(b)) = x$ then
      $a = b$. So $k_o h$ is a monomorphism.
(iii) h, k are isomorphisms if and only if they are epimorphisms and
      monomorphisms. By (i) and (ii) $k_o h$ is also both an epimorphism
      and a monomorphism, hence it is an isomorphism.
(iv) An epimorphism.
(v)   An epimorphism.
(vi) Monomorphisms in each case.

Exercise 9.12

This is a group of tallies with concatenation, together with their inverses,
which are isomorphic with the integers. See example 9.13.

Exercise 9.13

(i) $t^2 r$
(ii) r, t^2, rt^2, rt, rt^3
(iii) 8, for $rt^2 = t^2 r$, $rt = t^3 r$, $rt^3 = tr$

Exercise 9.14

No solution provided since the author is not a Rubik's cube addict.

Exercise 9.15

The set of integers which are multiples of N, with addition. If k, l are
multiples of N, so is $(k + l)$. '+' is associative, 0 the identity is a multiple
of N, and every multiple k of N has an inverse, $-k$, which is a multiple of
N.

Exercise 9.16

One should prove that h_T preserves the effect of the operations *Init, Add,
How many*. These are all based on addition of natural numbers with the
zero element, and the result follows from the fact that h_T is defined in
terms of summation.

Exercise 9.17

Each equivalence class comprises a set of cars which are the same in every respect except for their trim t.

Chapter 10

Exercise 10.1

> mods: $A^* \times \mathbb{N} \times A \to A^*$
> **pre**-mods$(l, n, a) \triangleq n \leqslant \text{len } l$
> **post**-mods$(l1, n, a, l2) \triangleq l2(n) = a$
> $\qquad \wedge \forall i \in \mathbb{N} . (i \leqslant \text{len } l1 \wedge i \neq n) \Rightarrow l2(i) = l1(i)$
> $\qquad \wedge \text{len } l2 = \text{len } l1$

> subs: $A^* \times \mathbb{N} \times \mathbb{N}_0 \to A^*$
> **pre**-subs$(l, n, k) \triangleq n + k \leqslant \text{len } l$
> **post**-subs$(l1, n, k, l2) \triangleq \text{len } l2 = k + 1$
> $\qquad \wedge \forall i \in \mathbb{N} . i \leqslant \text{len } l2 - 1 \Rightarrow l2(i+1) = l1(n+i)$

> front: $A^* \times \mathbb{N}_0 \to A^*$
> **pre**-front$(l, n) \triangleq n \leqslant \text{len } l$
> **posr**-front$(l1, n, l2) \triangleq \text{len } l2 = n \wedge \forall i \in \mathbb{N} . i \leqslant \text{len } l2 \Rightarrow l2(i) = l1(i)$

Exercise 10.2

(i)   *How many* $(b, c, e, t, Init( )) =$
How many $(b, c, e, t, [x \mapsto 0 | x \in (B \times C \times Eng \times T)])$
$\qquad$ by definition of the post-condition of *Init*

$= 0$

$\qquad$ by definition of the post-condition of *How many*

(ii)   From the post-condition of *How many*

How many$(b, c, e, t, Add(b, c, e, t, n, d)) =$
$Add(b, c, e, t, n, d)(\langle b, c, e, t \rangle)$

which, by the post-condition of *Add*

$= d(\langle b, c, e, t \rangle) + n$

which, by the post-condition of *How many*

$= $ *How many* $(b, c, e, t, d) + n$

(iii) The pre-condition of *Withdraw* is implied by the left-hand side of the implication because

$$How\ many\ (b, c, e, t, d) = d(\langle b, c, e, t\rangle)$$

We can therefore infer the post-condition of *Withdraw*, which is equivalent to

$$d(\langle b, c, e, t\rangle) = n + Withdraw(b, c, e, t, d, n)$$

which, applying the post-condition of *How many* twice, is equivalent to

$$How\ many\ (b, c, e, t, d) = n + How\ many$$
$$(b, c, e, t, Withdraw(b, c, e, t, d, n))$$

Exercise 10.3

(i)   **pre-**$GET(E, eid) \triangleq E(eid)$
         $GET(E, eid) \triangleq E(eid)$

or

         post-$GET(E, eid, inf) \triangleq inf = E(eid)$

(ii)   **pre-**$CHNAME(E, eid, name) \triangleq eid \in \text{dom } E$
         post-$CHNAME(E1, eid, name, E2) \triangleq$
         $\text{dom } E1 = \text{dom } E2 \wedge \forall e \in \text{dom } E1 \,.\, e \neq eid \Rightarrow E1(e) = E2(e)$
         $\wedge\ E2(eid) = \langle name, a, g, s\rangle$

where $E1(eid) = \langle n, a, g, s\rangle$

(iii)   **pre-**$CHAGE(E, eid, age) \triangleq eid \in \text{dom } E$
         post-$CHAGE(E1, eid, age, E2) \triangleq$
         $\text{dom } E1 = \text{dom } E2 \wedge \forall e \in \text{dom } E1 \,.\, e \neq eid \Rightarrow E1(e) = E2(e)$
         $\wedge\ E2(eid) = \langle n, age, g, s\rangle$

where $E1(eid) = \langle n, a, g, s\rangle$

(iv)   **pre-**$CHGRADE(E, eid, grade) \triangleq eid \in \text{dom } E$
         post-$CHGRADE(E1, eid, grade, E2) \triangleq$
         $\text{dom } E1 = \text{dom } E2 \wedge \forall e \in \text{dom } E1 \,.\, e \neq eid \Rightarrow E1(e) = E2(e)$
         $\wedge\ E2(eid) = \langle n, a, grade, s\rangle$

where $E2(eid) = \langle n, a, g, s\rangle$

(v)   **pre-**$CHSALARY(E, eid, sal) \triangleq eid \in \text{dom } E$
         post-$CHSALARY(E1, eid, sal, E2) \triangleq$
         $\text{dom } E1 = \text{dom } E2 \wedge \forall e \in \text{dom } E1 \,.\, e \neq eid \Rightarrow E1(e) = E2(e)$
         $\wedge\ E2(eid) = \langle n, a, g, sal\rangle$

where $E1(eid) = \langle n, a, g, s\rangle$

PART IV

Chapter 11

Exercise 11.1

$\forall a, b \in A . a R_R b \Rightarrow a R b \vee a = b$

Exercise 11.2

$\leqslant$

Exercise 11.3

a, c, d, e, f;

(g) R^{-1}, $\supseteq$, $\geqslant$, $\Rightarrow^{-1}$

(h)   Not in general, since R_f is not antisymmetric.

$\Rightarrow : R_f \times R_f \to \mathbb{B}$

where

$R_f \Rightarrow R_g \Leftrightarrow \forall a, b \in S . a R_f \Rightarrow a R_g b$

(i)   No, but its reflexive closure is.

Exercise 11.4

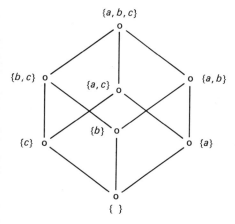

Exercise 11.5

(i)   c; g: $\geqslant$, $\supseteq$, $\Rightarrow^{-1}$ on $\mathbb{B}$.
(ii)   Yes.

Exercise 11.6

It should be antisymmetric, for otherwise one could have the physically impossible situation of a being a part of b and b being a part of a. Hence, 'is found in' is a partial order.

Exercise 11.7

(i)   To prove that $(a \sqcap b) \sqcap c = a \sqcap (b \sqcap c)$.

Proof:
By definition of $\sqcap$:

$$(a \sqcap b) \leqslant \{a, b\}; \text{ that is, } (a \sqcap b) \leqslant a \wedge (a \sqcap b) \leqslant b$$

also

$$x \leqslant \{a, b\} \Rightarrow x \leqslant (a \sqcap b)$$

Likewise

$$(a \sqcap b) \sqcap c \leqslant \{(a \sqcap b), c\}$$

that is

$$(a \sqcap b) \sqcap c \leqslant (a \sqcap b) \wedge (a \sqcap b) \sqcap c \leqslant c$$

but $(a \sqcap b) \leqslant a$ and $(a \sqcap b) \leqslant b$

so by transitivity

$$(a \sqcap b) \sqcap c \leqslant a, (a \sqcap b) \sqcap c \leqslant b$$

hence $(a \sqcap b) \sqcap c$ is a lower bound of $\{a, b, c\}$. Also

$$(x \leqslant (a \sqcap b) \wedge x \leqslant c) \Rightarrow x \leqslant (a \sqcap b) \sqcap c$$

but since

$$(x \leqslant a \wedge x \leqslant b) \Rightarrow x \leqslant (a \sqcap b)$$

we have

$$(x \leqslant a \wedge x \leqslant b \wedge x \leqslant c) \Rightarrow x \leqslant (a \sqcap b) \sqcap c$$

so

$(a \sqcap b) \sqcap c$ is the glb of $\{a, b, c\}$

By a similar argument

$a \sqcap (b \sqcap c)$ is the glb of $\{a, b, c\}$

But glbs are unique and so

$$(a \sqcap b) \sqcap c = a \sqcap (b \sqcap c)$$

(ii)   To prove that $a \sqcap b = b \sqcap a$.

Proof:

Immediate from the definition:

$$a \sqcap b \leqslant \{a, b\}$$
$$x \leqslant \{a, b\} \Rightarrow x \leqslant a \sqcap b$$

and the lack of ordering of the members of the set $\{a, b\}$.

(iii) The proofs of associativity and commutativity of $\sqcup$ are identical, since $\leqslant^{-1}$ is a po.

(iv)

Idempotency of $\sqcap$.

$$(a \sqcap a) \leqslant \{a, a\} = \{a\}$$
$$(a \sqcap a) \leqslant a$$

Also

$$x \leqslant a \Rightarrow x \leqslant \{a, a\} \Rightarrow x \leqslant (a \sqcap a)$$
$$a \leqslant a \Rightarrow a \leqslant (a \sqcap a)$$
$$a = (a \sqcap a) \text{ by antisymmetry of } \leqslant.$$

(v) Idempotency of $\sqcup$ follows since $\leqslant^{-1}$ is a po.

Exercise 11.8

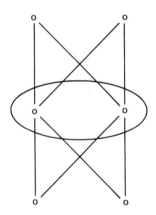

Exercise 11.9

(a) a, c, f, g: $\supseteq$, $\geqslant$, $\Rightarrow^{-1}$.
(b) Yes, Yes.
(c) No.
(d) No. There would have to be a part which was found in every other part (lower bound) and some large component in which every other part was found (upper bound).

Chapter 12

Exercise 12.1

(a) $y = 4$
(b) $y = 4 \wedge x * x = 9$

that is, $y = 4 \wedge (x = 3 \vee x = -3)$

(c) $y = -3 \vee x = 9$

Exercise 12.2

(a) $y = 4.$
(b)   False; that is, the condition is impossible.
(c) $x = 2.$
(d)   True; that is, the condition will always hold.

Exercise 12.4

(i) $x \geqslant 0 \wedge t > 0 \wedge (|r * r - x| \leqslant t \Rightarrow r = 4)$
that is, $x \geqslant 0 \wedge t > 0 \wedge |16 - x| \leqslant t \wedge r = 4$
(ii) $x \geqslant 0 \wedge t > 0 \wedge (|r * r - x| \leqslant t \Rightarrow (r \geqslant 3 \wedge r \leqslant 4))$
that is, $x \geqslant 0 \wedge t \geqslant 0$

$$\wedge ((r^2 \leqslant x + t \wedge r^2 \geqslant x - t) \Rightarrow (r \leqslant 4 \wedge r \geqslant 3))$$

that is, $x \geqslant 0 \wedge t \geqslant 0 \wedge x + t \leqslant 16 \wedge x - t \geqslant 9 \wedge r \geqslant 0$

In the first case the specification of square-root cannot *guarantee* the requirement R, although for any given x such that $x \geqslant 0$ and $|16 - x| \leqslant t$ there is an implementation of square-root which will satisfy R.

In the second case R can be guaranteed provided the implementation of square-root gives a positive rather than a negative answer.

Index

307